Arts-Based Methods in Education Around the World

RIVER PUBLISHERS SERIES IN INNOVATION AND CHANGE IN EDUCATION - CROSS-CULTURAL PERSPECTIVE

Indexing: All books published in this series are submitted to the Web of Science Book Citation Index (BkCI), to CrossRef and to Google Scholar.

Nowadays, educational institutions are being challenged as professional competences and expertise become progressively more complex. This is mainly because problems are more technology-bounded, unstable and ill-defined with the involvement of various integrated issues. Solving these problems requires interdisciplinary knowledge, collaboration skills, and innovative thinking, among other competences. In order to facilitate students with the competences expected in their future professions, educational institutions worldwide are implementing innovations and changes in many respects.

This book series includes a list of research projects that document innovation and change in education. The topics range from organizational change, curriculum design and innovation, and pedagogy development to the role of teaching staff in the change process, students' performance in the areas of not only academic scores, but also learning processes and skills development such as problem solving creativity, communication, and quality issues, among others. An inter- or cross-cultural perspective is studied in this book series that includes three layers. First, research contexts in these books include different countries/regions with various educational traditions, systems, and societal backgrounds in a global context. Second, the impact of professional and institutional cultures such as language, engineering, medicine and health, and teachers' education are also taken into consideration in these research projects. The third layer incorporates individual beliefs, perceptions, identity development and skills development in the learning processes, and inter-personal interaction and communication within the cultural contexts in the first two layers.

We strongly encourage you as an expert within this field to contribute with your research and help create an international awareness of this scientific subject.

For a list of other books in this series, visit www.riverpublishers.com

Arts-Based Methods in Education Around the World

Editors

Tatiana Chemi

Aalborg University
Denmark

Xiangyun Du

Qatar University
Qatar
and
Aalborg University
Denmark

River Publishers

Published, sold and distributed by:
River Publishers
Alsbjergvej 10
9260 Gistrup
Denmark

River Publishers
Lange Geer 44
2611 PW Delft
The Netherlands

Tel.: +45369953197
www.riverpublishers.com

ISBN: 978-87-93609-38-9 (Hardback)
978-87-93609-37-2 (Ebook)

Printed on acid-free paper.

Cover Picture credits to artist Okkyung Lee @jcslok

Contents

List of Contributors

Aimee Durning, *Faculty of Education, University of Cambridge, 184 Hills Road, Cambridge CB2 2PH, UK*

Arzu Mistry, *576, 1st Stage Indiranagar, Bangalore 560038, India*

Chiaki Ishiguro, *The University of Tokyo, Graduate School of Education, 7-3-1 Hongo, Bunkyo-ku, Tokyo 113-0033, Japan*

Dina Zoe Belluigi, *The School of Social Sciences, Education and Social Work, Queens University, University Road, Belfast, BT7 1NN, Northern Ireland*

Hiroaki Ishiguro, *Department of Education, College of Arts, Rikkyo University, 3-34-1 Nishi-Ikebukuro, Toshima-ku, Tokyo 171-8501, Japan*

James Biddulph, *Faculty of Education, University of Cambridge, 184 Hills Road, Cambridge CB2 2PH, UK*

Kimber Andrews, *University of Cincinnati, United States*

Kristóf Fenyvesi, *Faculty of Education, University of Cambridge, 184 Hills Road, Cambridge CB2 2PH, UK*

Luke Rolls, *Faculty of Education, University of Cambridge, 184 Hills Road, Cambridge CB2 2PH, UK*

Pamela Burnard, *Faculty of Education, University of Cambridge, 184 Hills Road, Cambridge CB2 2PH, UK*

Rannveig Björk Thorkelsdóttir, *School of Education, University of Iceland, Reykjavk, Iceland*

Susanne Jasilek, *Faculty of Education, University of Cambridge, 184 Hills Road, Cambridge CB2 2PH, UK*

Takeshi Okada, *The University of Tokyo, Graduate School of Education, 7-3-1 Hongo, Bunkyo-ku, Tokyo 113-0033, Japan*

Tatiana Chemi, *Department of Learning and Philosophy, Aalborg University, Kroghstræde 3, 9220 Aalborg, Denmark*

Tatjana Dragovic, *Faculty of Education, University of Cambridge, 184 Hills Road, Cambridge CB2 2PH, UK*

Todd Elkin, *365 Euclid Avenue, #103, Oakland, California 94610, USA*

Una MacGlone, *Reid School of Music, Alison House, 12 Nicolson Square, University of Edinburgh, Edinburgh, EH8 9DF, Scotland*

Xiangyun Du, *1. Department of Learning and Philosophy, Aalborg University, Kroghstræde 3, 9220 Aalborg, Denmark*
2. College of Education, Qatar University, Qatar

List of Figures

List of Tables

List of Abbreviations

HE	Higher Education
ILSP	Integrated Learning Specialist Program
ITA	Inspiration through art appreciation
M	Mean
MFA	Master of Fine Arts
NYC	New York City
SD	Standard deviation
SHoM	Studio Habits of Mind
SLR	Single-lens reflex
STF	Studio Thinking Framework
VR	Virtual reality
VTS	Visual Thinking Strategies

1

Arts-Based Methods in Education – A Global Perspective

Tatiana Chemi[1] and Xiangyun Du[1,2]

[1]Department of Learning and Philosophy, Aalborg University, Kroghstræde 3, 9220 Aalborg, Denmark
[2]College of Education, Qatar University, Qatar

Abstract

This chapter introduces the field of arts-based methods in education with a general theoretical perspective, reviewing the journey of learning in connection to the arts, and the contribution of the arts to societies from an educational perspective. Also presented is the rationale and structure of the book, as well as a summary of the following chapters.

1.1 Learning and the Arts: A Long Journey

To what extent the personal encounter with art and culture is important for the optimal development of children and young people has been repeatedly emphasised in theories of learning and development (Gardner 1994; Winner, Rosentiel & Gardner 1976) and by research findings (Deasy 2002; Fiske 1999; Rabkin & Redmond 2004). Research surrounding the encounters that children have with art and culture has been characterised by various approaches to child development, educational theories, and developmental psychology (Knowles & Cole 2008). Additionally, changing policies and strategies have influenced the research process (Akuno et al. 2015).

The present contribution was initiated by the intention of the editors to investigate these arts-based encounters from a global perspective. Being aware of the need for further studies that would touch upon the cultural, inter-cultural, cross-cultural, and global elements of arts-based methods in education, the editors started discussing a possible publication project in November 2016. The first call for papers was sent out in January 2017, and at the time the editors were expecting to collect about 12 contributions for a special issue that River Publishers was interested in hosting. The response from the scholarly community was overwhelming. The editors received such a large quantity of well-qualified, original, diverse, and relevant insights that the single publishing project multiplied. Against the background of this prolific harvest, the editors are offering the present special issue and wish to make the reader aware of two related works. The first is the anthology called *Arts-based Methods and Organisational Learning: Higher Education Around the World*, soon to be published for the Palgrave Series Studies in Business, Arts and Humanities, which focuses on higher educational settings and organisational learning (Chemi & Du, in press). The second project is a culturally specific anthology, which is currently being prepared with a projected publication date of 2018: *Arts-based Education- China and its Intersection with the World* (Du, Chemi & Wang, in preparation), a collection that focuses on cases from different regions of China, on Chinese art, and on cross-cultural projects involving China. This overwhelming response to the editors' invitation from different scholarly traditions and educational contexts is indicative of a growing global interest in arts-based methods in education. This tendency is consistent with a body of literature that has been based on high-quality research in recent years.

The most recent and exhaustive contributions that help frame the diversity and complexity of this field are *The Routledge International Handbook of the Arts and Education,* edited by Mike Fleming, Liora Bresler, and John O'Toole (2015); *The Routledge International Handbook of Creative Learning,* edited by Julian Sefton-Green, Pat Thomson, Ken Jones, and Liora Bresler (2011); the Waxmann's *International Yearbook for Research in Arts Education,* which has come to its third (edited by Shifra Schonmann in 2015) and fourth editions (this latest edited by Aud Berggraf Sæbø in 2016). In Britain, the living observatory of the programme Creativity, Culture & Education (2017) provides an always updated monitoring of creative partnerships involving the arts and education in a local and global perspective. However, already in 2007, Liora Bresler edited and published a fundamental contribution to the broad global perspectives on the field of arts education: the *International Handbook of Research in Arts Education.* In this extraordinary collection, the thematic sections touch upon multiple heated topics in the field of arts education studies and provide information about the global perspective of this field, thoroughly and systematically. Contributions and commentaries are collected from experts all over the world and involve a multiplicity of arts genres and traditions.

These contributions tend to bring together perspectives from all over the world and include a large variety of artistic genres and research methodologies. The topics that they touch upon span from policies to pedagogies, from social impact to philosophical conceptualisations. They are informative on specific topics, but they also offer a clear monitoring of the ways in which the general attention to the arts in education evolves through time. For instance, the introductory chapter in Fleming, Bresler, and O'Toole (2015) describes how policy statements have alternatively supported and ignored the needs of arts education. Here, and in the specific contribution of Akuno et al. (2015), is possible to read about the UNESCO attitudes towards this field as moving from reinforced priorities in 2006 and 2010, with respectively the Lisbon Conference (Portugal) and the South Korea World Conference, up to the surprising UNESCO rejection of the arts and creativity as priorities for education: "if nothing else, all that contemporary activity and global decision-making suggest a high level of both interest and confusion about the nature and the importance of the arts and their relationship to education" (Fleming, Bresler & O'Toole 2015, p. 1). Our contribution to the above literature aims at offering clarity on diverse practices and sustaining the theoretical and empirical attention to this field from a global perspective.

One example of the above-mentioned ebbs and flows in the field of arts education is the case of drama education in particular, where it is plain to see how educational practices are directly related to policy trends. When Bolton described the situation of drama education in the world in 2007, only a few years after the landscape had changed, it was modified according to fluctuating educational policies that chose whether or not to value the role of drama in education. According to Bolton (2007), the Scandinavian countries had a leading role in drama education going back to 1942. His review demonstrated how this positive trend continued through the first years of 2000, only to be substituted by a clear recession, which built up to the slow disappearance of drama as school subject or as a training subject at the educators' colleges. Very recently, Iceland inverted this trend by making drama education obligatory at all levels of compulsory schools (Thorkelsdóttir & Ragnarsdóttir 2016). According to Chemi (in press), "for a long time, Anglo-Saxon cultures seemed to have taken the lead in both research on drama education and in educational practices (Bolton 2007) as the Drama and Theatre section of the *International Yearbook for Research in Arts Education* in 2015 testimonies (Schonmann 2015)". In this way, is plain to see how research interests and perspectives follow the evolution of practices and policy. The case of drama education also brings forth the issue of cultural colonialism, which is as fundamental to global perspectives as ours and relevant to the arts at large. The assumed Anglo-Saxon expertise in drama education is criticised in Rajendran (2015) as part of a larger cultural exclusion: "Euro-American or Western centric interculturalism [. . .] tends to neglect histories and hierarchies of cultures" (p. 230). Despite this strong claim, the volume containing Rajendran's critique (the Drama and Theatre section of the *Yearbook* in Schonmann 2015) still disseminates cases drawn mostly from Western cultures or geographical placements. However, already in the subsequent yearbook, the 2016 *International Yearbook for Research in Arts Education* (Saebø 2016) reveals a more global picture, including examples of arts education from across different continents. The geography of arts education is slowly extending and reaching perspectives that are truly global and increasingly culturally inclusive. The present collection emerges from the need to continue making contributions in this direction, but further culturally diverse studies are still required.

Global trends are outlined in Wagner (2015), who considers the UNESCO research (Bamford 2006) and the OECD research (Winner & Vincent-Lancrin 2013) and Akuno et al. (2015) in this area to be the most recent and relevant landmarks in this field. Summarising the empirical contributions of these studies, Wagner (2015) emphasises a possible taxonomy, founded on "five basic approaches, paradigms, or objectives" (p. 25) to arts education: (1) the art-specific approach (artistic skills for their own sake); (2) the economic approach (economic output of creative industries); (3) the social approach (community projects with the arts); (4) the educational approach (integration of the arts in education); and finally (5) the political approach (building citizenship through the arts).

To the above categories, Chemi (in press) adds the health/therapy approach, which overlaps with several of the above-mentioned categories, but which she believes is an autonomous and independent perspective: "the application of the arts to health and therapy is a long-standing tradition and it has been made especially relevant to education through the self-regulation thinking. For instance, Sefton-Green et al. (2011) emphasise that the so-called *soft* skills of emotions regulation and of monitoring of cognition (meta-cognitive skills) are a fundamental part of the students' mental health and resilience" (Chemi in press).

We believe that global and cross-cultural perspectives are needed in order to fully understand what is really occurring in arts education and what the perspectives are for future practices, studies, and policies. In order to introduce the cultural and educational contexts in the current special issue, we wish to touch upon the changes in the theoretical conceptualisations that have characterised this field through the years.

1.1.1 The Arts Are Good for Learning

The constructivist approach to the child's self-development and self-expression has had a great impact on how the teacher and educator think about art, culture, and aesthetics as active learning tools in the classroom. Similarly, the role of Dewey's pragmatism (Jackson 1998) and his description of an experience-based pedagogy have stirred the theoretical understanding of the pedagogical role of the arts away from earlier, narrower directions. As Akuno et al. (2015) describe in their historical review, from the time of the ancient Greek conceptualisation of aesthetics, the arts seemed to deal primarily with beauty and morality (i.e. ethics and virtue).

Generally, from international research, it is confirmed that the involvement of art and culture in the lives of children does support their social, emotional, and cognitive well-being and development (Chemi 2014; Winner, Goldstein & Vincent-Lancrin 2013; Holst 2015). Whether this meeting is about being an active maker, or otherwise an (active) audience member, is not relevant, as both encounters activate complex responses and require engagement. Children's participation in artistic and cultural experiences appears to strengthen their ability to concentrate, and to engage in personal and social identity perception, which later in life may increase their confidence and ability to engage in social contexts. Within cultural and artistic experiences, children are offered a safe and challenging environment, resulting in a curiosity-stimulating, identity-building, and intellectually challenging approach that can lead to a positive attitude toward learning and development. The reason that the artistic learning environments are challenging is because they stimulate the child's ability to reflect, to find perspective, and to be critical, and they encourage creative thinking, empathy, and both metaphorical and logical-scientific thinking. Indeed, understanding art and cultural experiences is conducive to the overall development, whereby body, senses, cognition, and emotions are developed together. Artistic environments are often very safe despite any learning, understanding, and development challenges. The reason is that children who get early positive experiences with learning framed through arts and culture will have an opportunity to maintain a positive desire to develop further artistic experiences. This may contribute to learning readiness, creativity, and mental and emotional resilience later in life (Goleman 1995). Artistic learning environments address some of the children's prerequisites for learning: one learns through empirical, aesthetic, and discursive forms of learning, respectively (Austring & Sørensen 2006; Hohr & Pedersen 2001), and through their sensory language, arts and cultural experiences offer the opportunity to talk to both a real and a fantasy world—therefore addressing different learning approaches.

Ensuring accessibility to art and culture for school children means investing in community-building, and encouraging the development of resilient, robust, innovative, and competent individuals. This accessibility (Gardner 1994) is both physical and logistical (i.e. children actually have the opportunity to experience art and culture), but are also of mental and emotional nature (the cultural and artistic offerings must be developed with children, are meaningful in the child's life and important for their development). According to Perkins (1994), "bad habits of looking and thinking [that are] deeply rooted in the human organism" (p. 25) can be re-trained through positive habits, and the arts offer excellent cognitive training. The challenges hidden in the arts demand the activation of deep thinking, which happens to be rare in the shallow approach to art that is more typical in the mass culture of consumerism.

1.1.2 The Arts in Society

In recent years, no contribution to the arts in education seems to have been entirely free from the debate surrounding the role of the arts in education and in society at large. Advocacy can be present to a higher or lesser degree, but scholars and practitioners must often deal with questions about the impact of the arts on learning and society. No other culture beyond our Western-oriented, industrially advanced and technological culture under-prioritises the function of the arts in society, placing artistic practices in secondary roles compared to the production of goods, economic wealth, or commodity seeking. In Ward (2015), it is possible to find the historical reasons for this tendency, going back to Cartesian dualism (cognition *against* emotion) and the more mechanistic view on aesthetics formulated by Kant (Ward 2015, p. 108) and disseminated during the eighteenth century. If on one hand Kant had the fundamental role of acknowledging the human capacity for original thinking and creativity, on the other hand he understood aesthetic experiences as basically cognitive and almost subdued to mechanic reactions. Romanticism, in an attempt to transcend the perspectives on aesthetics produced during the Enlightenment period, ended up replicating the gap between reason and emotion, scientist and artist, Apollonian and Dionysian (Chemi, Jensen & Hersted 2015). The state of aesthetics during early and advanced industrialism is characterised by opposite tendencies, such as the dandy's "art for art's sake" as inconsistent to the instrumental use of the arts for acculturation's sake. Critical theories emphasise the risk that capitalism would instigate "new forms of oppression" (Ward 2015, p. 109). The consequences for the arts would be multiple: exclusion from power, the arts being seen as the arts "increasingly at odds with the material reality of life in the modern capitalist state" (Ward 2015, p. 109), or inclusion in the bourgeois *Bildung* as an element of cultural and moral elevation for the gentleman. Later on, the arts would transcend this gap in several ways, one of them being the closeness to progressive approaches to education, as the works of Dewey (2005) and Montessori (Ward 2015) confirm. Postmodernism made all the above contradictions explode in one rebel act against any restriction for the arts, with the consequence of either indulging in Marcusian pessimism about the emancipatory function of the arts, or experimenting with extreme, playful, and/or disruptive artistic behaviours. Examples of the latter can be the *murder* of authorial voice in the novel (as Barthes stated in *The Death of the Author* in 1967) or on stage (see Beckett's minimalistic plays, Chemi 2013) or in the visual arts (see pop art or street art). Today, neoliberalism and consumerism, together with "the lack of effective opposition to neoliberalism" from the artists' side (Ward 2015, p. 117), has turned the arts into a commodity to purchase or a service to measure. The value of the arts is measured in economic terms and put up against a monolithic practice of education. This (unfortunate) tendency contributes on one side to generate false expectations about the arts' educational output—either as a cure for all educational dysfunctions or as a mere distraction—and on the other side contributes to the construction of the rhetoric of advocacy. We hereby distinguish the discourses of advocacy from scientific arguments for the specific functions that the arts have always had—and still have—for human development, survival, and opportunities to thrive.

Arguments for the central role of the arts in human development and social life can be drawn from several disciplines and scientific traditions.

The neurosciences reveal the workings of the artistic brain (Levitin 2006) and how emotions are fundamental to learning (Immordino-Yang 2015; Immordino-Yang & Damasio 2007). Philosophy interrogates itself on the possible disruptive benefits that artistic practices can have on the imagination of future educational environments (Lewis 2012). In the studies of Ellen Dissanayake (2000, 1995), anthropology and evolutionary biology offer a clear link to the function of survival through social connectedness. According to these theories, what makes humans special is not the mere fact of living in societies, of being capable of making things (handiness), or of learning, but rather the fact that human beings are profoundly dependent on these activities. Many other animals live in clans; some of them even use tools (even though humans differentiate themselves for the tool specialization), and almost all of them learn from their upbringing, but only the survival of human beings strongly demands this. For this reason, humans had to come up with specific strategies, and art was one of these. Indeed, art came to permeate all elements and stages in life with an attitude that Dissanayake defines as "making special." Human beings are the only animals capable of and interested in cognitive abstractions such as symbolising, aesthetic elaboration, imagination, and innovation. The arts have been, through time and across cultures, inseparable from humanness and human development. In the current contribution, we will bring a variety of cases from different cultural and geographical contexts that describe in which ways the arts are—and will be for the future—fundamental tools and environments for education.

1.1.3 Our Contribution

The originality of our work in the present special issue relies on the variety of geographical contexts of the cases and writers' background (India, Japan, the US, the UK—including Scotland and Northern Ireland—Iceland, Denmark, Italy, and China), the novelty of empirical data, the variety of art forms (drama, improve, multi-arts, arts-integration, visual arts, movement, theatre, dance), and diversity of methodological approaches. When a collected work contains this level of diversity, however, the editors face a challenge in the effective presentation of a common thread. This is not because commonalities are not present, but rather because the possible angles to emphasise can be many. The potential stories can be endless. While discussing which angle to take, the editors became aware of an emerging dramaturgy, a sort of rhythm that the chapters were generating. This is the story that is going to be told and the reason for the chosen sequence of chapters.

In Aristotelian terms, a dramaturgical progression is constituted by: (1) introductory events; (2) a dramatic event or *catastrophe*; and (3) resolution of events, or *anagnorisis*. Following the Aristotelian template, Chapters 1–6 introduce the field of arts-based methods in education with **general theoretical perspectives** by Tatiana Chemi and Xiangyun Du in Chapter 1 and with **a list of cases** that touch upon **artistry in teaching** in Chapter 2 where Kimber Andrews explores the role of the teacher as a choreographer through an arts-based approach and how the teacher activates the curriculum through a performative approach in the classroom; **novelty and difference** in Chapter 3, where Tatiana Chemi addresses the topic of school learning enhanced and extended by means of artistic methods and approaches in the context of broader educational reform in Denmark; **enhancement of understanding** through in-depth analysis, using Activity Theory, of music improvisation activities for preschool children in Scotland by Una MacGlone in Chapter 4; **enhancing literacy** in Chapter 5 where Hiroaki Ishiguro depicts Japanese multimodal drama performance as child-centred performance ethnography in the form of a picture-mediated reflection on ‘Kamishibai’ (paper drama performance); and **arts-integration** in Chapter 6 by Todd Elkin and Arzu Mistry, through their studies on how accordion book practice supports the development of agency in teachers/learners/artists and promotes motivation and self-directed learning.

As the core of the whole collection, the very dramatic elements of arts-based methods *in the flesh* are represented by provoking and poetic contributions: in Chapter 7, Dina Zoe Belluigi shows how arts-based methods are well-placed to enable disruptions to the normative positioning of researcher, respondent and subject, drawing on the author’s reflections of opening the research processes to the possibilities of methodological ir/responsibility, which is **the bright and dark sides of this field**. In Chapter 8 Alison and colleagues present a set of experiential visual open work is built from a myriad of words, languages, cultures and critical theories, which is an attempt of **bringing the sensory (back) into scholarly research**, where it belongs.

Chapters 9–11 maintain the role of providing a resolution to the drama being told and offer perspectives for the further development of arts-based conceptualisations. These perspectives emphasise **the inspirational function of arts-based methods in education** in Chapter 9, where Chiaki Ishiguro and Takeshi Okada outline their model of the psychological process of inspiration through art appreciation (ITA), showing that its core consists of a dual focus on the artwork (and the artist) and the viewer’s own art-making, and **the enabling (and constraining) involved in educational encounters with the**

arts in Chapter 10 where Rannveig Björk Thorkelsdóttir tends to determine the aspects enabling or constraining the subject of drama in Icelandic compulsory education, using the lens of practice architecture theory. Finally, in Chapter 11, with the purpose of bringing this dramaturgical storytelling to a conclusion, Pamela Burnard and colleagues suggest **arts-based methods as the creation of possibilities for future education**.

References

Akuno, E., Klepacki, L., Lin, M. C., O'Toole, J., Reihana, T., Wagner, E., and Restrepo, G. Z. (2015). "Whose arts education?," in *The Routledge International Handbook of the Arts and Education,* eds M. Fleming, L. Bresler, and J. O'Toole. (New York, NY: Routledge), 79–105.

Austring, B. D., and Sørensen, M. (2006). *Æstetik og læring*. Copenhagen: Hans Reitzels Forlag.

Bamford, A. (2006). *The Wow Factor: Global Research Compendium on the Impact of the Arts in Education.* New York, NY: Waxmann.

Berger, R. (2003). *An Ethic of Excellence: Building a Culture of Craftsmanship with Students*. London: Heinemann.

Bolton, G. (2007). "A history of drama education: a search for substance," in *International Handbook of Research in Arts Education* L. Bresler (Dordrecht: Springer), 45–62.

Booth, D. and Gallagher, K. (2003). *How Theatre Educates: Convergences and Counterpoints with Artists, Scholars and Advocates*. Toronto: University of Toronto Press.

Bresler, L. (Ed.). (2007). *International Handbook of Research in Arts Education*. Dordrecht: Springer.

Catterall, J. S., Chapleau, R. and Iwanaga, J. (1999). "Involvement in the arts and human development: general involvement and intensive involvement in music and theatre arts," in *Champions of Change: The Impact of the Arts on Learning,* Ed. Fiske, E. B (Dordrecht: Springer), 1–18.

Creativity, Culture and Education. (2017). *Research Reports*. Newcastle Upon Tyne: CCE.

Charyton et al. (2009). creativity as an attribute of positive psychology: the impact of positive and negative affect on the creative personality. *J. Creat. Mental Health*, 4, 57–66.

Chemi, T. (2012). *Kunsten at Integrere Kunst i Undervisningen*. Aalborg: Aalborg Universitetsforlag.

Chemi, T. (2013). *In the Beginning Was the Pun: Comedy and Humour in Samuel Beckett's Theatre.* Aalborg: Aalborg University Press.

Chemi, T. (2017). *A Theatre Laboratory Approach to Pedagogy and Creativity: Odin Teatret and Group Learning*. Basingstoke: Palgrave.

Chemi, T. and Du, X. (Eds.). (2017). *Arts-based Methods and Organisational Learning: Higher Education around the World.* Basingstoke: Palgrave Studies in Business, Arts and Humanities.

Chemi, T., Jensen, J. B. and Hersted, L. (2015). *Behind the Scenes of Artistic Creativity: Processes of Learning, Creating and Organising*. Frankfurt: Peter Lang.

Csikszentmihalyi, M. (1996). *Creativity: Flow and the Psychology of Discovery and Invention.* London: HarperCollins.

Deasy, R. J. (Ed.). (2002). *Critical Links: Learning in the Arts and Student Academic and Social Development.* Washington, DC: Arts Education Partnership.

Dewey, J. (2005). *Art as Experience*. London: Perigee.

Dissanayake, E. (1995). *Homo Aestheticus. Where Art Comes from and Why.* Seattle, DC: University of Washington Press.

Dissanayake, Ellen. (2000). *Art and Intimacy: How the Art Began.* Seattle and London: University of Washington Press.

Du, X., Chemi, T. and Wang, L. (Eds.). (2018). *Art-based Education- China and its intersection with the world.* Basingstoke: Palgrave.

Eisner, E. W. (2002). *The Arts and the Creation of Mind.* New Haven, CT: Yale University Press.

Fiske, E. B. (Ed.). (1999). *Champions of Change: The Impact of the Arts on Learning*. Available at: http://artsedge.kennedy-center.org/champions/pdfs/ChampsReport.pdf

Fleming, M., Bresler, L., and O'Toole, J. (Eds.). (2015). *The Routledge International Handbook of the Arts and Education.* New York, NY: Routledge.

Gardner, H. (1994). *Frames of Mind. The Theory of Multiple Intelligences.* London: HarperCollins.

Goldstein, T. (2010). *The Effects of Acting Training on Theory of Mind, Empathy, and Emotion Regulation*. Ph.D. dissertation, The Graduate School of Arts and Sciences, Boston, MA.

Goleman, D. (1995). *Emotional intelligence, why it can matter more than IQ*. New York City, NY: Bantam Books.

Hetland, L., Winner, E., Veenema, S., and Sheridan, K. (2014). *Studio thinking 2*. New York, NY: Teachers College Press.

Hohr, H., and Pedersen, K. (2001). *Perspektiver på Æstetiske Læreprocesser (Perspectives on Aesthetic Learning Processes).* Frederiksberg: Dansklærerforeningens Forlag.

Immordino-Yang, M. H. (2015). *Emotions, Learning, and the Brain: Exploring the Educational Implications of Affective Neuroscience (The Norton Series on the Social Neuroscience of Education).* New York City, NY: WW Norton & Company.

Immordino-Yang, M. H. and Damasio, A. (2007). We feel, therefore we learn: the relevance of affective and social neuroscience to education. *J. Compil.* 1, 3–10.

Jackson, P. (1998). *John Dewey and the Lessons of Art.* New Haven, CT: Yale University Press.

Journal of Aesthetic Education. (2000). Special Issue. The Arts and Academic Achievement: What the Evidence Shows. Arts Educ. Pol. Rev. 34, 2–90.

Knowles, J. G., and Cole, A. L. (2008). *Handbook of the Arts in Qualitative Research: Perspectives, Methodologies, Examples, and Issues.* Newcastle upon Tyne: Sage.

Levitin, D. J. (2006). *This is Your Brain on Music: The Science of a Human Obsession.* Chennai: Penguin.

Lewis, T. E. (2012). *The Aesthetics of Education: Theatre, Curiosity, and Politics in the Work of Jacques Rancière and Paulo Freire.* New York, NY: Bloomsbury Publishing USA.

Marshall, J. (2005). Connecting art, learning, and creativity: a case for curriculum integration. *Stud. Art Educ.* 46, 227–241.

McLellan, R. et al. (2012). *The Impact of Creative Initiatives on Wellbeing: A Literature Review.* Newcastle upon Tyne: Creativity, Culture and Education.

Perkins, D. N. (1994). *The Intelligent Eye: Learning to Think by Looking at Art.* Los Angeles, CA: Paul Getty Trust.

Perkins, D. N. (2009). *Making Learning Whole: How Seven Principles of Teaching Can Transform Education.* San Francisco, CA: Jossey Bass.

Rabkin, N. and Redmond, R. (Eds.). (2004). *Putting the Arts in the Picture: Reframing Education in the 21st Century.* Chicago, IL: Columbia College Chicago.

Rajendran, C. (2015). "Intercultural drama pedagogy: dialogical spaces for empathy and ethics," in *International Yearbook for Research in Arts Education 3/2015: The Wisdom of the Many- – Key Issues in Arts Education,* Ed. S. Schonmann Munster, NY: Waxmann Verlag, 229–234.

Reimer, B. (1992). "What knowledge is of most worth in the arts?," in *The Arts, Education, and Aesthetic Knowing,* eds B. Reimer and R. A. Smith (Chicago, IL: The University of Chicago Press), 20–50.

Reimer, B. and Smith A. R. (Eds.). (1992). *The Arts, Education, and Aesthetic Knowing*. Chicago, IL: University of Chicago Press.

Robinson, K. (2001). *Out of Our Minds: Learning to be Creative*. Chichester: Capstone.

Tishman, S. and Palmer, P. (Eds). (2007). "Works of art are good things to think about," in *Evaluating the Impact of Arts and Cultural Education,* (Paris: Centre Pompidou), 89–101.

Thorkelsdóttir, R. B. and Ragnarsdóttir, Á. H. (2016). "Analysing the arts in the national curriculum in compulsory education in iceland," in *International Yearbook for Research in Arts Education 4/2016: At the Crossroads of Arts and Cultural Education: Queries meet Assumptions*, A. B. Sæbø (Münster, NY: Waxmann Verlag), 179–188.

Schonmann, S. (Ed.). (2015). *International Yearbook for Research in Arts Education 3/2015: The Wisdom of the Many-Key Issues in Arts Education*. Münster, NY: Waxmann Verlag.

Sefton-Green, J., Thomson, P., Jones, K., and Bresler, L. (Eds.). (2011). *The Routledge international handbook of creative learning*. New York, NY: Routledge.

Sæbø, A. B. (Ed.). (2016). *International Yearbook for Research in Arts Education 4/2016: At the Crossroads of Arts and Cultural Education: Queries meet Assumptions*. Münster, NY: Waxmann Verlag.

Vecchi, V. (2010). *Art and Creativity in Reggio Emilia: Exploring the Role and Potential of Ateliers in Early Childwood Education*. New York, NY: Routledge.

Wagner, E. (2015). Local-Global Concepts in Arts Education. in *International Yearbook for Research in Arts Education 3/2015: The Wisdom of the Many-Key Issues in Arts Education,* ED. S. Schonmann (Münster, NY: Waxmann Verlag), 24–29.

Wakeford, M. (2004). "A short look at a long past," in *Putting the Arts in the Picture: Reframing Education in the 21st Century,* eds Rabkin, N. and Redmond, R (Chicago, IL: Columbia College Chicago), 81–106.

Ward, S. C. (2015). "The role of the arts in society," in *The Routledge International Handbook of the Arts and Education,* eds M. Fleming, L. Bresler and J. O'Toole (New York, NY: Routledge), 106–121.

Winner, E. (1982). *Invented Worlds: The Psychology of the Arts*. Cambridge: Harvard University Press.

Winner, E., Hetland, L., Veenema, S., Sheridan, K., Palmer, P., and Locher, I. (2006). Studio thinking: how visual arts teaching can promote disciplined habits of mind. *New Dir. Aesthet. Creat. Arts*, 189–205.
Winner, E., Rosenstiel, A. K., and Gardner, H. (1976). The development of metaphoric understanding. *Dev. Psychol.* 12, 289.

2

Artistry in Teaching: A Choreographic Approach to Studying the Performative Dimensions of Teaching

Kimber Andrews

University of Cincinnati, United States

Abstract

In this study, the role of the teacher as a choreographer of the educational experience is explored in relation to how they make the curriculum come alive in the performative space of the classroom. Using an arts-based approach, I applied concepts and tools from the study of dance to analyse the embodied dimension of teaching. I examine how two university professors vary the pacing, energy, and focus of the class to communicate concepts and ideas to the students, as well interweave multiple streams of information through their embodied communication and the narrative structure of the class.

2.1 Introduction

In Anne Bogart's (2007) book *And Then You Act: Making Art in An Unpredictable World* she writes, 'The translation of page to stage is the translation of the logic of ideas and words into the logic of time and space' (p. 12). The elements of time and space are inherent to the performing arts. They are the canvas directors, choreographers, filmmakers, and performance artists create upon. Teaching is a performing art in the sense that it is both temporal and spatial. The classroom is the stage, and the teacher performs or enacts their curriculum in front of a student audience. Some may stand at the podium reading from lecture notes, others move around, gesture to power points, call on students, but all must translate the goals, objectives, and lessons on the syllabus into the performative realm of time and space.

In this chapter, I explore the embodied and performative approaches of two university teachers, and how they orchestrate the curriculum in the classroom. I take an arts-based approach that applies tools and techniques used in the study of dance to analyse the movements and sequential structure of a lecture style classroom. This study is focused on what Eisner (2002) terms the 'curriculum in vivo,' or the actual activities employed in the classroom and the translation of the 'intended curriculum or the curriculum in vitro' into an embodied and sequential narrative that brings the curriculum to life. Using a choreographic framework, I illuminate the aesthetic and embodied dimensions of teaching.

2.2 Artistry in Teaching

As Rubin (1985) points out, 'The research on pedagogy describes the competencies and characteristics of skilful teachers. It tells us *what* should be done, but not *how* it should be done. ... The research on artistry ... tells us a bit more about the how' (p. 91, italics original). The 'how' or qualitative and embodied dimension of teaching can be difficult to describe because of the ephemeral nature and the challenges of finding language that can encapsulate the complexity of experience. Polanyi (1966) characterizes this as the tacit dimension—what we know but cannot say (p. 4). However performing artists, particularly in theatre and dance, develop tools for studying the tacit through extensive training that cultivates sensitivity to how one's tone of voice, posture, and bodily stature imbue a narrative with meaning. For instance, the way a dancer circles the shoulder could be read as a coy gesture to tease the audience, or a shrug of indifference.

Expert teachers also use the tacit and performative in the classroom. Eisner (2002) writes, 'Judgment depends on feel, and feel depends on a kind of somatic knowledge.... The body is engaged, the source of information is visceral, the sensitivities are employed to secure experience that makes it possible to render a judgment and act upon it' (p. 201). He points to the way artists attune to the relationships between parts, and they do so not through a strictly analytic engagement of ideas, but also with the visceral and somatic. Actors, dancers, and musicians actively cultivate an embodied awareness to make judgments on how to react to qualitative stimulus with great sensitivity. One of the characteristics of artistic teachers is their ability to react to and improvise on their class plans in order to meet the needs of the students—whether that be changing the agenda for the day because students are not understanding fundamental concepts needed to move forward, or choosing to stand motionless at the front of the classroom in order to quiet students rather than shouting to get their attention.

Performers and teachers share the broader goal of communicating to an audience, but also must help students and viewers connect to material. Sarason (1999) is concerned that education,

> . . . glosses over the nature and complexity of the phenomenology of performing: How and why it requires a teacher to think, feel, intuit and flexibly adapt to students' individuality, and to do all of this for the purpose of engendering understanding and as a sense of growth. When we say that performers seek both to instruct and *move* an audience, we mean that the teacher as performing artist has in some positive way altered the students' conception of the relationship between sense of self and the significance of the subject matter, i.e., an increase in competence. (p. 48, italics original)

Sarason points to the lack of emphasis in education on the relational and performative nature of teaching and makes an important point about the necessity for teachers to *move* their students towards greater understanding—move indicating making the material personally relate to the student audience. Sarason makes vivid teachers need to think about how to present material in a way that engages students in self-growth and discovery, rather than merely covering concepts in the syllabus.

One of the major critiques of focusing on the performance of teaching in education is that it runs the risk of becoming 'edutainment,' or information presented in an entertaining way that lacks true substance. Pineau (1994) fears 'the claim that teaching is performance will evoke nothing beyond the facile acknowledgement that a certain theatricality can help hold the attention of drowsy undergraduates in early morning or late afternoon classes' (p. 5). She suggests one of the reasons the field of education has been critical of an artistic or performance approach to education is that it has largely been based on comparing the role of a teacher to an actor. She describes the actor-cantered approach as 'Performance is reduced to style, and further, to a particular style of enthusiastic theatricality employed to energize one's communicative behaviours' (p. 6).

Pineua's critique is apt given some of the literature that aims to develop the performance styles of teachers is laden with prescriptive advice about how to use humour, dramatic readings, and role-playing to boost enthusiasm in the classroom. Tauber and Mester's (2007) book *Acting Lessons for Teachers* indentifies enthusiasm as the most important quality for teachers to express and gives a variety of examples on how a teacher might spice up their classroom performance. This type of discussion fails to recognize the wide variety of teaching styles that may not emphasize enthusiasm but be quite successful at engaging students. If taken at face value, books of this sort can simplify the role of performance in teaching to that of a cheerleader, rather than acknowledging the complex nature of both good performance and teaching. It accentuates the exterior veneer of performance, without a discussion of the context, audience interaction, and development of self that rigorous performance training entails.

Louis Rubin (1985) in his book *Artistry and Teaching* looks specifically at what artistic teachers do. He provides many descriptions and criteria for artistic teachers such as, flair, originality, craftsmanship, discerning judgment, and extraordinary perception, but boils it down to 'artistry implies human accomplishment that is unusual in its proficiency and cleverness, and greatly superior to conventional practice' (Rubin, 1985, p. 16). Rubin acknowledges there is not one approach to artful and effective teaching. Although the skills common to effective teachers has been explored and articulated in the literature, not all teachers exemplify all of the characteristics on the list. He writes,

> Teachers do specific things to accomplish their goals. It is not acting, per se, nor salesmanship, nor communication, nor entertainment, nor pedagogical jugglery which account for performance. It is the gestalt of these—moulded into a personal style, built around individual attributes, and energized by genuine commitment and an educated mind—which account for teaching that takes students beyond the confines of their interests. (Rubin, 1985, p. 163)

My impetus for studying college educators is to examine the gestalt nature of artistry in teaching and how various layers of communication assemble to imbue the classroom with meaning. As the authors discussed above, performance is an often-overlooked aspect of teaching and poses distinct challenges to study because of its affective nature. In this study, my aim was to explore how tools and techniques used to study movement and choreography in dance could be used to analyse the embodied dimension of teaching and articulate this often underexplored or articulated aspect of teaching.

2.3 Choreography as a Framework for Exploring Teaching

This study focuses on the *how* of teaching—primarily how each teacher enacted the curriculum in the classroom. In other words, I wasn't so interested in the *what*, or the content of the class, but was more invested in looking at how teachers structured the curriculum and imbued it with meaning through their performance. Eisner (2002) reiterates the relationship between course content and how it taught when he writes,

> *How* one teaches something is constituent with *what* is taught. Method or approach infuses and modifies the content that is being provided. Thus, teaching becomes a part of curricular process, and curricular processes, including their content, become part of teaching; you can't teach nothing to someone. (p. 150, italics original)

Although the disciplinary culture and content of each course plays a role in teaching, I have focused on the temporal structure of the course—how a teacher organizes activities within the time allotted for their class—rather than the actual content of the course. In other words, what Boogart (2007) calls the *translation* of the curriculum into the live performance space of the classroom.

Having spent my early career as a professional dancer and choreographer, I drew extensively on that experience as I moved into teaching in lecture style classrooms. I found designing curriculum and orchestrating it in the classroom had many similarities to creating dances. In this study, I was interested in looking at classroom teaching through the lens of choreography to illuminate the relationship between embodied communication and the structure of the class.

Using a choreographic framework to look at teaching highlighted three key insights into the artistry of teaching. First, teachers and choreographers both create structures from scratch. Unlike in theatre where a director typically begins with a script, a choreographer, like a teacher, designs the overarching themes and narratives of the class. Second, choreographers oversee the entire aesthetic of a performance. Although many think a choreographer's primarily role is creating movement sequences, they also select sound/music, lighting, costumes, etc. that support their artistic vision and create the ambience for the dance to flourish within. Similarly, teachers create a wide variety of content for a course—assignments, exams, lectures, etc.—yet *how* they deliver that content and orchestrate learning in the classroom contributes to creating an environment for learning. Third, a choreographic framework accentuates the role of the body in teaching. The basic elements of choreography are space (the area in which the body moves), time (the speed or pace of how the body moves), and energy (the quality of the movement). The moving body is a central component of choreography and using it as a framework for looking at teaching provided a language to articulate and analyze the role that movement plays in the classroom.

A choreographic framework allowed me to explore the structure of class—how each teacher orchestrated the unfolding of time and the narrative structure of the class—in relation to their embodied performance of the material. This helped me distinguish good performance from the curriculum. To use an analogy from theatre, it enabled me to determine the difference between a good play mired with bad acting, and conversely, an underdeveloped plot saved by the excellence of sensitive actors.

The choreographic framework also had limitations. Focusing on the performative dimension of teaching did not generate findings on the quality of the content of the class, nor if the skilful performance of the teachers increased student learning. The findings from this study focus on articulating the often overlooked embodied dimensions of teaching to bring to the forefront what was tacitly being communicated through the teacher's performance.

2.3.1 Data Collection

Stake (1995) suggests 'The researcher should have a connoisseur's appetite for the best persons, places, and occasions. "Best" usually means those that best help us understand the case, whether typical or not' (p. 56). I observed several professors before selecting two participants, Julie and Alexandra.[1] This allowed me to identify three criteria to evaluate if a teacher was a potential candidate for the study. First, the teacher must have a dynamic range as a performer and use the body as part of their instruction. For instance, I observed two professors that gestured often, moved around the room and spoke loudly, but these had little relation to what they were actually communicating. Both professors attempted to engage students with an enthusiastic performance, yet it created a type of monotone that was loud and energetic but lacking variation.

[1]The names of the professors selected for this study have been changed to provide anonymity.

Second, I needed to have a working understanding of the content covered in the class to see the relationship between the curriculum and how the body was used as a tool for communicating ideas and concepts. I observed a professor of microbiology who was a dynamic performer, but the advanced nature of the course and content made it difficult to understanding the major concepts covered in the class. Although I was not explicitly studying the content of the course, it was important that I could grasp the general concepts and follow the narrative structure of the course.

And third, it was important the teachers selected for the study had different teaching styles to add complexity and texture to the case. Julie and Alexandra had two distinctive styles of performing in their teaching. Alexandra is a professor of finance and has been teaching for over 20 years. She is the course coordinator for an upper level finance class that all undergraduate students must pass to graduate. She and her colleague co-authored the textbook for the class, and she regularly revises the course content and teaching strategies to meet the needs of the evolving student population.

Julie is a professor of public relations and has been teaching at the university for 9 years. I observed her teach an advanced public relations course. Most of the students in the class were advertising majors, and the course was not required for graduation but a popular elective. Julie teaches a variety of classes in her department and this was the second time she had taught advanced public relations.

I selected these two instructors because they provided a distinct contrast in teaching styles in the classroom, and each attuned me to different issues of how performance is used in the classroom. While there was little similarity between the courses, departments, and contexts of the classes studied, this study is not meant to compare and contrast particular ways of teaching a specified class, but instead to look at how teachers design, orchestrate, and use the body as a tool for communicating in teaching. In many ways, the different contexts for each class—the size, the students, and departmental contexts—added to the complexity and richness of the study, as well as highlighting different approaches to orchestrating the curriculum in the classroom.

I attended Alexandra's finance class and Julie's public relations class for the entire 15-week semester with the exception of exam days, and collected three kinds of data for this study: observational field notes, semi-structured interviews, and documents from the classes studied (syllabi, assignment prompts, etc.). In addition, I interviewed Julie and Alexandra three times each during the semester for approximately one hour, and conducted an additional interview of 1.5 hours at the end of the semester. All interviews were audio recorded and transcribed. I also had numerous conversations before and after class with both teachers and documented these as closely as I could from memory in my field notes. The majority of my questions in interviews with Julie and Alexandra were based on specific observations from the class. By asking questions based on my observations, I was able to gain an understanding of the motives, philosophies, and rationales behind activities, policies, and the way they altered the structure or performance during the class.

2.3.2 Embodied Method

Observing the way each teacher used her body to communicate in the classroom was essential for this study. In a previous study of dancers in a choreography class, I began to devise a method for translating embodied observations into written language. I further developed this method into a 4-part process using tools and techniques from studying movement in dance to analyse the teachers embodied personas through a sensory approach. In phase one of the process, I studied the teachers movements through an embodied approach (dwelling and interiorization), and in phase two I worked to articulate embodied observations into language that could render embodied observations visible to a reader (transcription and translation).

Polanyi (1966) describes the process of learning embodied knowledge as moving from dwelling into interiorization. In the first phase, a student *dwells* with an expert observing and mimicking movements until it becomes natural in their own body, and they have *interiorized* it. The process Polanyi describes articulates how dancers study and learn movements. First they observe a teacher or choreographer do a sequence of movements. Next they begin to co-dwell observing and trying to do the movement in their own body. Through reflective repetitions of the movement sequence, a dancer will eventually get a sense the rhythm, pacing and weighted quality of the movements and interiorize the sequence into their own body. Homans (2010) uses a quote from ballerina Natalia Makarova to illustrate this process of interiorization. Markarova described the process of interiorization as, 'dancers are trained . . . to "eat" dances—to ingest them and make them part of who they are.' (Homans, 2010, p. xix)

Studying the movement of Alexandra and Julie provided a distinct challenge; I could not co-dwell with the teachers in the classroom as I would in a dance class. Instead, I would take notes trying to capture the details of the gestures and movements I observed. Then I would go home and recreate these movements and gestures as closely as possible in my own body. This process of dwelling was iterative and with each new observation, I refined the details of my embodied performance. However, I found to interiorize the quality of the movement—the rhythm, weighted quality, and energetic dynamic—I needed to *watch with body*. This meant not taking notes, quieting the analytic mind, and letting my body absorb the affective qualities of each teacher's movement. I would very subtly allow my body to move and respond to the teacher's movements and later would conjure the felt sensations when reconstructing movements, gestures and postures observed in class. I was not able to fully interiorize the quality of each teacher's movements without periods of engaging my kinaesthetic awareness and knowledge.

In phase two, I began to articulate my embodied observations into words that attempted to capture what I had observed. In the first phase, I wrote detailed transcriptions of common gestures breaking down each movement and describing it in words. Although these were helpful in meticulously analysing the movement, they were tedious to write and even more tedious to read. The fourth phase of the process was translating these transcriptions into language that captured the affective dimension and what it communicated in the classroom. Rather than trying to transcribe what I was observing, I attempted to write descriptions that encapsulated the feeling, tone, and overall affect of the movement within the context of the situation, and often this led me to use poetic imagery and metaphor to help the reader visualize what I had observed.

The 4-part process of translating embodied observations into written language was time-consuming, but it gave me a unique insight into the experiential dimension of embodiment that I have not found through any other method. The point of translating embodied observations into written language is not to say that I understand the participant's reasons for moving or can know their feelings by doing the movement in a similar way (Stinson, 2004), but rather, they allow for a rich description that is based on what the body was actually doing. For instance, instead of saying the teacher looked displeased before class started, by taking on the posture of the teacher, I can describe the action. Instead, I might say the teacher pushed her weight into the podium, leaning forward, as her eyes intensely scanned notes before class.

It is also important to note that neither Alexandra nor Julie was purposely using movements and gestures in the class. In fact, I would show them common gestures in interviews and ask them to comment on their perceptions of what they might be communicating. They often would reply, 'I do that?'

Their gestures and movements were fully integrated into their teaching persona; hence my job was to bring the embodied into the foreground in order to examine the tacit nature of embodiment in teaching.

2.4 A Day in the Life of the Classroom

The concept of teaching style relates to what Eisner (1979) identifies as the implicit curriculum or what is conveyed through the activities, structure, tone of voice, disposition, and the emphasis a teacher puts on certain aspects of class. In the following section, I will highlight vignettes from the first day of both Alexandra and Julie's class. First impressions are important in teaching. The first day provides valuable information on what the course will be about, but also the general attitude and disposition of the teacher. If one thinks of the first day like a movie trailer—a short synopsis to give a potential audience member an idea of what the film is about—both teachers gave an accurate representation of the aesthetic of their teaching style on the first day of class. The structures of Julie and Alexandra's first day presentations were similar in form, but not in style or delivery. I will move back and forth between the two to accentuate the contrast.

2.4.1 Before the Class Begins

Alexandra sits casually leaning back in a tall office chair watching a live stream of Bloomberg Television[2] on the computer monitor while the students see it on the projector screen. On the bottom half of the screen is a window with the international date and time counting up the hours, minutes, seconds, and right now it is 9:20:12 a.m. A handful of students, six or seven, are sitting near the front of the classroom at tables that seat two; some are watching the projection of Bloomberg, others are looking at their phones. Alexandra sits at the podium/technology cabinet on the right hand side of the room. She occasionally looks at her own phone, and smiles at a funny commercial. More students pour in until the class is almost full—about forty students distributed in five columns of tables that seat two. The room is relatively quiet; the students face forward bracing for what is to come. This finance course is notorious because it is difficult and students need to pass it before they can take other upper level courses.

[2]Bloomberg Television is an United States based international news station that delivers business and market news.

At 9:29:30 a.m. on the international time at date clock, Alexandra stands, takes off her jacket and places it on the back of the chair, and at precisely 9:30:00 a.m. she hits a button on the computer and the sounds of reporters squawking and ticker tapes running stops abruptly. The website for the class appears on the screen and class begins.

*****

At 11:52 a.m. the lecture hall is full of chatter. Students stream in and greet people they know with smiles, waves and hellos. The seats are partially filled with clumps of three and four students sitting together taking up about half of the approximately 200 seats. The hall is long and narrow 12 seats across and 17 rows back. There are no windows and wood panelling lines the walls partway up. At 11:55 a.m. the door at the front of the lecture hall swings open and in walks Julie. She smiles, sets her stuff down on a table at the front of the room and scans the crowd. She gives little waves and says hellos to students she recognizes, and then sits on the table at the front of the auditorium and begins to swing her legs as she continues to scan the class. The room is full of chatter, but Julie catches the attention of the students in the first three rows when she asks if anyone went anywhere exciting over winter break. A student responds that she went to Mexico and Julie responds, 'I love Mexico, but my husband never wants to go because he says it is too dangerous.' She then leans forward to confide to a few students in the front row that she is 'rockin' the hat' because she burnt her hair with a straightening iron and will be wearing hats until she can get it fixed.

She looks at her watch at 12:00 p.m., and says, 'I'll give it a few more minutes,' meaning she will wait to start class, because students might be late on the first day. At 12:02 p.m. she says, 'Alrighty, let's get started' as she pushes herself off the desk and moves to the centre of the room.

2.4.2 During the Class

At 9:30 a.m., the projector screen that once had Bloomberg Television playing now displays the course calendar. Every class session is mapped out with titles like 'Financial Markets' and the accompanying Power Point can be clicked on to download. Deadlines for tests, group projects, and homework are noted on the schedule, and Alexandra carefully goes through each requirement: weekly homework, three midterms and a cumulative final, three group projects, and pop quizzes. The workload is daunting, and Alexandra tells the students that this class is 'Finance through a fire hose, because the fun just keeps on coming.'

After systematically going through the calendar and basic guidelines for the class, Alexandra switches modes, and begins a section of the lecture she calls 'eccentricities.' She explains that all teachers have expectations and pet peeves that are unique, and she wants to be up front about hers. First, she explains that she doesn't accept late work and emphatically tells students that assignments need to be put in the mailbox on her office door, not in the general mailbox in the business building where 'anyone has access to it.' Second, she doesn't negotiate grades; they get what they earned. Alexandra then continues and gives them strategies for doing well in class. She suggests that students read the chapter before coming to class, and do the practice questions at the back of the book after class. Alexandra makes the distinction between 'knowing' something and being able to 'demonstrate' that you can do it, and suggests studying is not reading the book, but applying that knowledge through doing practice questions.

She ends her discussion on eccentricities by saying, 'Class starts at 9:30. If you are late, please come in through the back door, so you don't disturb your fellow classmates.' She also says, 'if want to do something else, like surf the web or text on your phones, go someplace else to do it' because 'it hurts my feelings when you are not paying attention to me.' She says this with an exaggerated frown mocking sadness. She ends by saying, 'In finance, the fun never stops, so let's get started,' and she begins a lecture on strategies for investing that takes up the final 45 minutes of class.

*****

After getting off of the table, Julie takes centre stage at the front of the auditorium and begins the class by asking, 'How many of you have had a public relations course before,' and many students raise their hand. She asks a few more questions about what courses they have taken and then says, 'Okay, what do you guys want to know?' She pauses for a second while looking out at the sea of students, and with a sly smile says, 'No, I will tell you' indicating that she will go over the guidelines for the class.

But she doesn't tell them just yet; she continues to ask them questions—what kinds of jobs do people in public relations do, and the definition of key terms, like what is a 'public' and the difference between a 'strategy' and a 'goal.' She is quick to call on students as hands go up and leans on the first row of theatre seats in the direction of the person talking.

She then asks again, 'Okay so what else do you want to know?' She pauses, looks upwards, and nods her head as if checking off things she needs to discuss. Then, she tells the students there will be two group projects—one based on building a P.R. campaign for clients she has recruited, and the second is a case study presentation. She advises students to email her if they want to be in specific groups, and she will assign the rest of the students randomly.

Julie says, 'Okay, what else' and pulls out a piece of paper from her folder, scans it quickly, and announces, 'Most of this stuff is on the class site,' but does not use the projector to show the students the class website. She then explains that if the weather is really bad, she might not be able to make it into town because she lives in the country, but that rarely happens, but to check their email before class on bad weather days.

She asks students if there is anything they really want to cover. No one responds at first, and then a student says she wants to learn about the different agencies they could potentially work for, and Julie says she would be happy to Skype in alums and have them discuss their first jobs. She tells them she is 'happy to bring people in and do what interests them as long as we can fit it into the schedule.'

She reiterates that everything they need—syllabus, group assignment prompts, etc.—are on the website, and they can look at it later. She then switches modes and says, 'So here are some things you should know about me. I start out liking all of you. Some people say I have "pets" but really the only thing you have to do is come up and talk to me, and I will like you.' She summarizes, 'basically, don't harsh my mellow.' Julie reiterates that she likes all of her students, but says, 'If I catch you cheating, I will bust you' and her tone lowers and one eyebrow goes up as she looks out. Again she states, 'Really guys, don't harsh my mellow, and we will be fine.' Julie ends by telling the students that 'you would have to be a total looser to fail this class,' and that if they come to class and do the work, they will do well.

2.4.3 Summary

First impressions are just that—impressions. The first class is an important window into what is to follow, but certainly over the course of 15 weeks things can and do change. Yet, a student in Alexandra's class would likely surmise that timeliness is important, and the due dates are fixed; she is no nonsense, the class will move at a quick pace, and each class will cover a substantial amount of material. Julie seems to be approachable, easy going, and interested in getting to know the students and engaging them in conversations before and during class. The first class was focused on finding out what the students already knew about public relations and putting them at ease about the class—the information is online, and if the students try, they should have no problem passing the class.

2.5 The Choreography and the Dance: Curriculum Embodied

In this section, I will discuss the choreography of the classroom, or how each teacher approached sequencing and structuring the activities in the class. Drawing on observations over the entire semester and interviews with both participants, I develop two frameworks to describe Alexandra and Julie's approach to planning classes. Next, I will discuss the embodied dimension of teaching and how each teacher's physical performance adds an additional layer of meaning in the classroom.

2.5.1 Scripting an Experience

I would characterize Alexandra's approach to teaching as a scriptwriter because of the time she invests and the emphasis she puts on structure in the course. When observing Alexandra teach, you have the feeling that nothing was left to chance. Each lecture feels like a calculated performance that includes stories, detailed Power Point slide presentations, and mathematical problems worked out in class to see if students understand the concepts. I had the opportunity to observe a few lectures twice (once in the Fall semester during the pilot study, and then again in the Spring), and although there was some variation, the lectures included the same stories to exemplify a concept. Watching Alexandra teach was akin to seeing an experienced actress play a role she has done many times, but yet, still finds ways to make it feel fresh each time.

Alexandra often begins class with a story or metaphor. For instance, when discussing portfolio diversification she asked, 'Why is it bad to put all of my eggs in one basket.' She mimed having a basket resting on her forearm and directing students to look into the imaginary basket to visualize the idea of having a basket full of things as she walked up one of the isles. She used this metaphor as the basis for the rest of the lesson on diversifying investments. Later in the semester, during the NCAA college basketball tournament, she asked students to bet on which team would win. She engaged the students in thinking through how someone could trust that if they put $100 on a team, they would get their money. She went on to describe the role of a bookie in betting, and then related this to the futures market and the way a clearinghouse functions in a similar way.

Alexandra comes from a long line of storytellers. Her grandfather was a newspaper editor in the Netherlands. During World War II, he was known for cleverly using the Dutch language to give the Nazi propaganda he was forced to publish a double meaning. Alexandra said that storytelling and writing is a part of her family linage, and that it is common in her family to speak in parables. She described,

> Whenever we want to get a point across, we always tell a parable. When you are talking to people you tell them straight, but you give them a story, so they understand where it fits. And we've always been [storytellers] in our household. It's true, when we are trying to get a point across, we give the parable because it is much easer to say, 'I get that because I see it,' rather than I understand the abstract nature of it.

Alexandra's classes have a theatrical quality, not only because of her energetic performance, but also because of the tight narrative. Alexandra doesn't meander through topics, and she doesn't mince words. Her lectures have the quality of a script that was carefully written, and then enacted with sensitivity to the audience's reaction. In a student interview, when I asked what characteristics he would use to describe Alexandra's teaching, he replied, 'She doesn't waste words.' It is clear from the years of teaching this course that her performance is well-rehearsed.

2.5.2 Creating Experiences on the Fly

Julie's style of teaching was more spontaneous and improvisational. I often wrote in my field notes, 'fly by the seat of the pants,' to describe her approach which, at times, was exhilarating. For instance, on the third day of class, she dedicated the entire session to leading a discussion on a burgeoning issue at the university. It was a subzero temperature day and many colleges in the surrounding area had cancelled class, but this particular university did not and sent out an email advising students to be cautious getting to classes. Students began a twitter hashtag to respond to this, and some sexist and racially disparaging tweets about the chancellor were included. This incident made national news. Julie spent the entire class leading a discussion on how the university could deal with this from a public relations standpoint. A student in the class was featured in a *Huffington Post* article two days later giving his take on the issue through a tweet. Julie recognized this publically by projecting the article on the screen before class and applauded him when he entered the classroom.

Julie often began class with a public relations issue that was currently in the news. When she didn't have an example prepared, she would invite students to share issues related to public relations and on the spot pull up a video online related to the topic to watch and discuss. The success of these discussions varied greatly depending on the examples chosen and the student's interest and willingness to participate.

Julie did lecture and used detailed Power Point slides that were essentially outlines from chapters in the assigned textbook. Julie often illustrated concepts from the textbook with public relations cases that served as practical examples, or she would make up scenarios to illustrate these concepts. There was always a sense of spontaneity to Julie's lectures over the textbook material. While lecturing, she typically turned her back to the students to look up at the Power Point slide with her index finger on her chin—a stature of thinking. She would quietly read the text to herself under her breath and nod as she took the information in. By the time Julie turned around, she had thought of an example related to the concepts on the Power Point. For instance, one day she was talking about non-violent protests and said, 'An example of this would be like if the Girl Scouts had a sit-in at the rotunda of the state capital on a day they were going to talk about STEM education to make them [legislators] aware of the lack of women in science.' She stopped and smiled delighted with the thought, and said in a quieter voice, 'That would actually be pretty cool wouldn't it' as she directed her gaze to a few of the students sitting in the front of the class.

I asked Julie how she prepares for the course, and she said that she creates all of the Power Points before the semester begins. She said, 'Pretty much, the foundation is ready to go at the beginning of the semester, so that I can do fun things or refine it during the semester.' I then asked about her daily preparation, or how she thinks about making the material come alive in class. She replied,

> Some of them [lectures] I have done enough times, I know what I am going to say. If there is a current event going on, I can draw that in. I have some examples on my slides, but I usually come up with those on the fly. Some days are better than others. Sometimes we

> have these case studies [student presentations on a public relations case]. I mean, usually there is something that we are talking about in the material and I can bring in something in from that [case study presentations] into the discussion.

She conceded that she likes to be flexible with her class plan, and that this 'drives some student's crazy,' but reiterated that this is the nature of public relations—things change quickly, and one must adapt.

2.5.3 Structure Embodied

In this section, I will explore the embodied dynamics of Alexandra and Julie's teaching, and how these added another layer of communication to the classroom choreography. The above characterizations of Alexandra as a scripter and Julie as an improviser were visible in the ways they used their body's to communicate the content of the class. For instance, Alexandra's movements tended to accentuate what she was saying—her circling arm gesture signalled the forward movement needed to think through a problem, and when her arm outstretched to the power point screen it was to highlight a specific line of text or point on the graph not a general gesture in that direction. The precision of her scripted text was also mirrored in the specificity of her gestures.

Julie's performance was less didactic and rhythmic than Alexandra's. One movement flowed into the next as the words cascaded out of her mouth. She occasionally used specific gestures to accentuate what she is talking about—like drawing her hands together in front of the body to signal getting 'focused'—but usually, the gestures were more an indicator of her own bodily attitude at the moment. For instance, sometimes she stood with wide legs in the centre of the classroom when she wanted to demand attention, or put a finger to her lips as she listened to a student and thought about what they are saying.

A helpful analogy to describe the aesthetic uniqueness of Julie and Alexandra is music because it encapsulates the tonal quality of the voice in coordination with the energetic qualities of the movement. In general, Julie's performance was more melodic and Alexandra's percussive.

Julie projected her voice without being loud. Her words came out with feathery edges that smooth one sound to the next, but each sound was clear and articulate. There was something calming about her voice. It was soft without being meek, steady but never booming. The rise and fall of the tone of her voice reverberated through the room.

Julie's postures were typically casual; she sat on the backs of chairs with her feet on the seat facing the students, leaned against the wall, and occasionally sat on the floor in the isles during student presentations. There was a shuffle to her walk as she meandered up one isle with her 'Mom' mug in hand. Julie attempted to simulate the feeling of an intimate conversation in a large lecture hall with her students. She was informal and invited conversation, just as she would in a one-on-one situation. The major difference was the sphere of her attention. Julie was able to expand it in a large classroom to embrace the entire room, and with ease, zero in on a student that was talking, and then widen it back out to the rest of the room. Her presence in the classroom was embracive.

In contrast, Alexandra's performance was dramatic, punctuated, and had a pronounced rhythm. She hurtled words out of her mouth making sure to hit every consonant—tongue against teeth. Rather than a streaming melody, Alexandra spoke in cadences that shifted from progressive rhythmic explorations, to sudden silence, and an occasional boom of the base drum. Her voice was full of tension, getting higher in pitch as she got louder—a shrill trill when she was excited or exasperated. She used her gestures to illustrate concepts like widening the hands away from one another to show the market is growing and bringing them together to show it is shrinking. Finance lends itself to didactic gestures—markets move: up and down, grow and shrink, crash—and Alexandra illustrated these movements with her body.

Alexandra made use of dramatic pacing in her speaking. She knew how to pause and dangle suspicion or explode into a faster pace. The body often reflected this pacing; at times, she stood in a pause position, held her hands behind her back talking through a concept, and then glided over to the podium with a finger dangling in the air right before letting it fall down dramatically to set the Power Point into motion. Her pacing was significant. The pauses signalled a space for students to process information before moving on, creating an ebb and flow of information with breaks in between. This was especially apparent to me as a non-finance/math person. Although I often did not understand the mathematical equations, nor could I do the problems assigned in class, I usually understood the concepts discussed.

Both Alexandra and Julie have great range—they could be explosive, quiet, measured, casual, but these took on different qualities in each teacher. Alexandra was animated, had stories to tell, and appeared to fully relish in divulging the information bit by bit; the class unfolded so students see the beauty of markets, numbers, and Excel documents. The finance class was her stage and the performance had been thoroughly structured and rehearsed, but there is never a feeling that she was merely 'going through the motions.'

Each performance was fresh and at 9:30 a.m. the curtain opened, and Alexandra was rip-roaring ready to go.

Julie was more like a talk-show host or the ringleader of the class. She was clearly the star of the show. She directed the conversations, posed questions, set up situations for discussion, but the class felt improvised—the content of what was being discussed was perhaps less important than keeping the discussion going. She attempted to draw students out by asking them questions like, 'what do you think about this' inviting them to personally respond and this often took the class away from conversations based in public relations and into other discussions related to the students' beliefs and opinions. She spoke the students' language regularly using slang words like chick, dude, lol, etc. Although Julie related to the students as peer in her speech and attitude, she was clearly the host of the 'show' in class responsible for calling on students and responding/redirecting the conversation.

Leder (1990) describes 'the absent presence' (p. 3) of the body in everyday life and many scholars on embodiment refer to the embodied dimension as the 'background' (Radman, 2012). I found that the embodied dimension of teaching similarly functions on a subterranean level. The structure of the class provided a narrative, yet the embodied performance of each teacher acted like a perfume that permeated the class. The percussive quality of Alexandra's performance and tight narrative created an atmosphere where one needed to sit up and be ready to go. Students often arrived a few minutes before the class began to review notes and prepare for the lecture. The more melodic and casual quality of Julie's performance created a relaxed environment and students sometimes would whisper quietly to one another during lectures.

Yeat's (1989) poses the question in his poem *Among School Children*, 'How can we know the dancer from the dance?' The embodied dimension of teaching is intertwined with the choreographic structure of the class just as the dancer and the dance itself are difficult to distinguish. My interest in making a space to look at the choreography and embodied dimension of teaching individually has been to highlight the often overlooked tacit forms of communication that add layers of meaning that are difficult to discern and describe.

2.6 Conclusion

Dewey (1934) characterizes an aesthetic experience as leaving an audience with a feeling of 'consummation rather than cessation' (p. 37). In an aesthetic experience, there is a seamless flow of ideas that is continuous and all parts of the experience are integral to unifying the whole. In regards to teaching, it is the difference between covering key concepts in order to check them off the list, to weaving the same concepts together to tell a broader story of their relevance.

Artistry in teaching is complex because there are many variables to contend with: the curriculum, the students, the classroom space, the time of day the class is offered, the teachers energy level, and sometimes even the weather makes a difference. Finding ways to use all of these elements in the classroom to engage students in learning is a form of choreography, and on the good days, everything seems to move towards consummation. Eisner (2002) points out the connection between the curriculum design and how the teacher in the classroom expresses it. He writes,

> How one designs a lesson or curriculum unit matters, and the design of such plans and activates depends every bit as much on attention to relationships among their components. In the course of teaching matters of pacing, timing, tone, direction, the need for exemplification are components whose relationships need to be taken into account. The ability to do so constitutes a part of the artistry inherent in excellent teaching. (Eisner, 2002, p. 202)

Looking at the classroom as a work of art through a choreographic frame allowed me to see the variations between Julie and Alexandra's approach to structuring the course and how they communicated and implemented that plan in the class.

This study also reiterates Pineau's concern that an actor-centered approach to the performance of teaching is flawed. Artistry in teaching comes from an integrated approach that includes the design, implementation, and classroom performance of teachers. Rubin suggests that there is not one way to become an artistic teacher, but in this case study it is apparent that a thoughtful design and ability to synthesize concepts so the students see how the materials of the class coalesce is impetrative. An engaging performance is not enough, and it takes a thoughtful course design as well as finding ways to translate that plan into the time and space of the classroom.

Barone and Eisner offer 'arts based research is the effort to extend upon the limiting constraints of discursive communication in order to express meanings that would otherwise be ineffable' (Barone & Eisner, 2012, p. 1). In this study, I have developed an embodied method and framework based on approaches dancers use to study movement to illuminate the embodied dimension of teaching. My work is in response to writers like

O'Loughlin (2016) who registers concern that if the embodied or tacit dimension of learning is not addressed then we run the risk of instilling actions and reactions to others unconsciously. She writes,

> These values and categories also encompass activities undertaken to form corporeal habits but end up being much more than corporeal habits. Since their entry into the individual is not by means of the presentation of ideas and concepts, but instead by means of direct bodily intervention, they in fact bypass consciousness, becoming ingrained as basic orientations towards the world. A cognitive paradigm unfortunately denies the body's active intentional capacities. (O'Loughlin, 2006, p. 69)

In my attempt to translate the ineffable tacit dimension of communication into a discursive language, my desire is provide a tools for analysis from the arts to better understand the complexities of the performative and embodied aspects of teaching.

References

Barone, T. & Eisner, E. (2012). *Arts Based Research*. Thousand Oaks, CA: Sage.

Bogart, A. (2007). *And then, You Act: Making Art in an Unpredictable World*. New York, NY: Routledge.

Dewey, J. (1934). *Art as Experience,* 2005th Edn. New York, NY: Penguin Group.

Eisner, E. (1979). *The Educational Imagination: On the Design and Evaluation of School Programs*. New York, NY: Macmillan Publishing Company.

Eisner, E. (2002). *Arts and the Creation of Mind*. New Haven, CT: Yale University Press.

Homans, J. (2010). *Apollo's Angels: A History of Ballet*. New York, NY: Random House.

Leder, D. (1990). *The Absent Body*. Chicago, IL: University of Chicago Press.

O'Loughlin, M. (2006). *Embodiment and Education: Exploring Creatural Existence*. Dordrecht: Springer.

Pineau, E. L. (1994). Teaching as performance: Reconceptualizing a problematic metaphor. *Am. Educ. Res. J.* 31, 3–25.

Polanyi, M. (1966). *The Tacit Dimension,* 1st Edn. Garden City, NY: Doubleday.

Radman, Z. (ed.) (2012). *Knowing Without Thinking: Mind, Action, Cognition and the Phenomenon of the Background.* Hampshire, NY: Palgrave Macmillan, 187–205.

Rubin, L. J. (1985). *Artistry in Teaching,* 1st Edn. New York, NY: Random House.

Stake, R. E. (1995). *The Art of Case Study Research.* Thousand Oaks, CA: Sage Publications.

Stinson, S. W. (2004). "My body/myself: Lessons from dance education," in *Knowing Bodies, Moving Minds: Towards Embodied Teaching and Learning,* ed. L. Bresler (Dordrecht: Kluwer Academic Publishers), 153–167.

Tauber, R. T., and Mester, C. S. (2007). *Acting Lessons for Teachers: Using Performance Skills in the Classroom,* 2nd Edn. Westport, CT: Praeger.

Yeats, W. B. (1989). *The Collected Poems of W. B. Yeats*. London: Wordsworth Poetry Library.

3

New and Different: Student Participation in Artist-School Partnerships[1]

Tatiana Chemi

Department of Learning and Philosophy, Aalborg University, Kroghstræde 3, 9220 Aalborg, Denmark

Abstract

The present contribution addresses the topic of school learning enhanced and extended by means of artistic methods and approaches. In the context of broader school reform in Denmark, new opportunities emerged for schools to partner with cultural institutions that are external to schools but informally involved in learning processes. Among a variety of external partners–such as sports clubs, entrepreneurs, cultural clubs-artists and cultural institutions were chosen as the focus of the research project Culture Laboratory. The qualitative study that documented and assessed the nine artist-school partnerships within Culture Laboratory showed a large number of learning outcomes, as reported by the main participants (students, teachers, artists, cultural institutions). This chapter will first of all describe the general context and purpose of the overall research project. Secondly, it will elaborate on a specific set of findings, which demonstrate a particular emotional response in the students' experience: emotional arousal. This response is described as surprising, exciting, novel and different, and brings with it a number of learning outputs.

[1]The present contribution is an elaborated and revised version of a research report published in Danish (Chemi 2017b). The translation from Danish is mine.

3.1 Something New in the State of Denmark

In 2014, the Danish school system had to deal with the introduction of a brand new school reform. Amongst the several changes outlined, one in particular has provided schools with an unexpected opportunity to collaborate with cultural institutions and artists: the Open School. This change consisted in a clear expansion of the schools' cultural framework through activities and projects that could more systematically bring art and culture to schools and schools to cultural institutions. This action is not just about a new organisation of teaching, but rather about a whole new view of learning and art, and therefore about new dilemmas. As early as 2015, the national association of Danish municipalities, KL (2015), published a guide to the Open School, but the publication did not specifically focus on cultural or artistic partnerships, instead applying a broader understanding of partnership as collaboration with sports institutions, clubs and entrepreneurs. The school reform, which was adopted in 2013 and came into force in August 2014, demanded forms of cooperation that opened the school to the surrounding local community. According to that part of the reform called the Open School, schools were not only expected but explicitly requested to work more systematically with local organisations such as sports associations, business or entrepreneurial environments, after-school cultural projects or so-called culture-schools (*kulturskoler*). This aimed to create for primary schools "a framework for experience, immersion and enlightenment, so that the students would develop knowledge and imagination and gain confidence in their own opportunities and a background for action" (Ministry of Education 2017, my translation from Danish). The relevance of this case lies in the novelty of the political initiative, making it possible to investigate the early steps of formalised artist-school partnerships, in a socio-cultural context that has valued the arts as learning opportunities over an extended period of time.

In my previous research (Chemi 2014, 2015) I have noticed that, in collaborative practices between artists and schools, uncertainty affects teachers and educators as well as artists. Teachers are often unsure of what students should learn, how to organise their schedule, what their role should be in partnerships with artists and how to evaluate children's learning. Even though they value the arts as experiential and pedagogical tools, they still might not feel comfortable in introducing the arts throughout the curriculum. On the artists' part, uncertainties are most commonly about their role in the partnership and about their art: does art and creation drown in learning goals and frameworks? Is art to be exploited for other purposes than the artistic?

Should the artist be a teacher? Previous research projects (Chemi 2014) have documented what happens when art is integrated into teaching and how schools can create good learning environments by focusing on artistic, aesthetic and creative methods. On the one hand, the projects show that there is great benefit in making more room for the arts in school; on the other hand, a specific challenge is thrown up: the cultural encounter between the ways of thinking of schools and artists. These approaches can be so different that partnerships can be impossible to carry on. Will, commitment, passion for the arts must be present for both parties. The problem is that the two parties are often ignorant of each other: many practitioners like the idea of a different school that contains the aesthetic, bodily, emotional and sensory elements, but how are they supposed to achieve that?

The standard curriculum of the *practical-musical* subjects (as defined in the Danish *Folkeskole*) seems unambitious, with its narrow focus on sports, visual arts/design and music. Compared to Anglo-Saxon countries where one can find as compulsory subjects drama/performance and dance, both Danish *Folkeskole* and teacher education are far from offering a wide range of basic artistic skills. For example, drama in Denmark is often covered by other school subjects, such as Danish or foreign languages, and dance is usually included in gymnastics, sport and movement. Artist-school partnerships or integration of art in interdisciplinary education are mostly left to the individual teacher's will and competencies, often obtained through hobbies rather than teacher education. Collaboration with artists can be experienced as difficult because, among other things, teachers often lack skills in what I define as arts integration (Chemi 2014), and which is an internationally recognised practice (Marshall & Donahue 2014). Arts-integrated projects that aim at enhancing the quality of participation in artistic experiences and at the incorporation of different forms of art into the schools' core curriculum appear to be a kind of add-on in Denmark. Teachers who are willing to prioritise experiential or experimental teaching are often *brokers* for arts-integrated projects. The arts-integrated activities are often project-based and collaborative, make use of group work and interdisciplinary relations, and engage the students' own strengths, their senses, well-being and emotions. Arts-integrated projects can be designed against the background of a wide variety of art forms and teaching methods and are experienced by artists, children and teachers as both challenging and engaging.

In the context offered by the 2014 school reform and the opportunities defined by the Open School, I carried out a study that aimed to document and evaluate a broad development project that took place in Funen (*Fyn*) between

January 1, 2016 and June 30, 2017. The present contribution sums up selected findings from the research publication that, so far, has been published only in Danish (Chemi 2017b) and that disseminates the multiplicity and wide reach of artistic partnerships with schools.

3.2 A Laboratory for the Arts and Culture

The development project called Culture Laboratory (*Kulturens Laboratorium*) was originally designed by a team of visionaries in the region of Funen. From autumn 2014, an on-going conceptualisation around the project's aims and possibilities took the form of dialogue between the region and myself as a researcher, invited to take part as an expert in arts-based learning and informal sparring partner. It was decided that the development project should cover nine partnerships that would be formed between a school, an artist and a cultural institution, and each party's contribution would be documented qualitatively. Research data were to be gathered from the nine partnerships, from April 2016 to December 2016. During this period, artistic activities were implemented, documented, mapped and analysed.

The Culture Laboratory has become reality within the economic, organisational and content framework that the Cultural Region of Funen has made available. According to Danish policy measures, a cultural region is "one or more municipalities that have concluded a cultural agreement with the Ministry of Culture" (Ministry of Culture 2017, my translation from Danish). Funen's agreement focused on dialogue between art, culture and school as educational institutions. This meant that political backing was explicit and not merely symbolic: actions were taken against the background of a growing awareness around the positive role of the arts for learning.

Several arguments provided a solid background for the project. At international level, UNICEF's children's rights declaration stipulates that all children have the right to art and cultural experiences, as clearly stated in Article 31: "Children have the right to relax and play, and to join in a wide range of cultural, artistic and other recreational activities" (UNICEF 1989). Similar arguments are described in the Strategy for Young Children's Encounter with Arts and Culture (Ministry of Culture 2014), where the need to look into children's involvement in cultural and artistic communities was determined and made public. The Ministry of Culture's strategy for the encounter of children and young people with art and culture, launched in May 2014, called for a qualitative study of how aesthetics are part of children's everyday lives – specifically, with a study of artist-school partnerships. The Culture

Laboratory project was established to meet this requirement. The project's purpose was to investigate qualitative trends in the use of art and culture in schools. In addition, the project aimed at getting closer to good examples that could inspire regional and national future experiences, and at getting acquainted with the strengths and dilemmas in the integration of arts and culture in schools. The research project offered a qualitative survey based on the nine selected cases of school cooperation with the arts and culture.

The originality of the project lies in the multifaceted perspective that accommodates all learning perspectives: students, teachers, artists and cultural institutions. Diversity is also expressed in the research design, which covers several geographical areas in the region, multiple art forms, different school stages (early years – K9, intermediate years – about 10–12 years old, senior years – about 13–16 years old), and various activities and organisations. The research study was based on previous experiences with similar evaluations of art and culture (Chemi 2014), mapping cultural offers in learning environments, pedagogical research using quantitative as well as collaborative data and the possibility of developing educational inspirational materials. Unlike other contributions on partnerships and the Open School, the current contribution aimed at adding to the qualitative diversity by highlighting the participants' reactions, by looking closely at the specific elements of the artistic and cultural activities, and by focusing on the general and specific prerequisites for learning and the special contribution of arts and culture to learning outputs.

3.3 Partnership: What's in a Name?

The Danish National Network of School Services (Nationalt Netværk af Skoletjenester 2015, 2016a, 2016b) has been the initiator of several mappings within the Open School area and artists' or cultural institutions' partnerships with schools. Each contribution has had a particular focus on a specific dilemma, such as transportation between school and cultural institutions, or schools' use of educational facilities and partnerships. The 2016 mapping (Nationalt Netværk af Skoletjenester 2016a) has contributed both qualitatively and quantitatively to the knowledge of various Danish initiatives and experiences. That research publication also disseminated useful resources that are continuously made available by municipalities in order to facilitate the workings of partnerships.

In the 2016 publications we also find a variety of definitions of the word *partnership* (2016a, pp. 13–16). The Culture Laboratory project did not explicitly define what a partnership is or should be and the interpretation was, in the end, left to the concrete practices of sub-projects. It is thus possible to observe that a general understanding of the concept of partnership has emerged from practice. Here, a partnership was seen as a binding and equal collaboration among three parties – a school, an artist, a cultural institution – whose purpose was to conceive, design and carry out engaging educational programmes for children in primary school, based on the children's encounter with the arts and professional artists, as well as with culture and professional cultural agents.

3.4 The Arts Education Tradition

The theoretical framework behind this study is the now-established tradition of arts and arts-based education, which is outlined in the introductory chapter of the present special issue. Its core concept is Gardner's pivotal idea of *participation in the arts*, which owes to Nelson Goodman's (1976) theory of symbols, where the arts are seen as complex and engaging understanding. According to Gardner, "participation in the artistic process" (Gardner 1994a, p. xii) is an accessibility of both physical and logistic nature (children having the concrete opportunity to experience art and culture), but also of mental and emotional nature: the cultural and artistic offerings are developed for, with and by children and are meaningful in the child's life and significant for the child's development. The concept of active engagement in the arts is influenced by Gardner's pluralistic understanding of the child's intelligence (Gardner 1994b). Gardner seems to suggest, on one hand, a broader look at participation in the arts, which can be based on active making or active appreciation, and on the other hand, a wider understanding of intelligence that includes the aesthetic, sensory and bodily dimensions. According to Gardner (1994a), when children have artistic experiences they participate actively in a process that is *per se* educational, because "the arts create expectations and then resolve or violate them, and by doing so they stimulate complex intellectual responses that integrate both affection and cognition" (Chemi, Jensen & Hersted 2015, p. 97). Moreover, art and cultural subjects offer an aesthetic form of recognition that characterises creative processes (Gardner 1993, Csikszentmihalyi 1996). This form of recognition can lead to self-esteem and flow experiences (Csikszentmihalyi 1990), which in turn can lead to industrial or social innovations. By ensuring the accessibility of art and culture for school children, society invests in robust, innovative and competent people.

These arguments are recurring in the scholarly tradition of arts and arts-based education studies (Bresler 2007, Fleming, Bresler & John O'Toole 2015, Sefton-Green, Thomson, Jones & Bresler 2011, Schonmann 2015, Berggraf Sæbø 2016) and can be the basis of advocacy for the arts (Bamford 2006), of research-based argumentation for learning outputs (Deasy 2002; Fiske 1999; Rabkin & Redmond 2004) or of sceptical investigation of the inner workings of participation in the arts in relation to learning outcomes (Winner, Goldstein & Vincent-Lancrin 2013). Amongst the large-scale studies that inspired the present research was that of Davies et al. (2013) which evaluated *Curriculum for Excellence*, initiated by the Scottish Government through Learning and Teaching Scotland. They found reasonable grounds for establishing that activities taking children and young people out of school to work in cultural learning environments, such as museums and galleries, improved the students' creative abilities.

The constructivist awareness of the child's self-development and self-expression is what in Denmark has most influenced how teachers and teacher education understand arts, culture and aesthetics as active learning tools. Nevertheless, this area still needs systematic study, as in Denmark we do not know enough about the involvement of art and culture in our schools. A good start in this direction was made by Bamford (2006) as part of her global investigation for UNESCO and as part of her specific look at Denmark, requested by the then Minister of Culture (Bamford & Qvortrup 2006). However, these investigations were not taken further by sustained research, even though their findings led – in direct and indirect ways – to the establishment of the Open School initiative. Today, the Open School throws up new demands for in-depth documentation, as teachers, artists and cultural institutions might not feel prepared to enter into partnerships as learning environments. When the art is involved in specific activities with educational objectives, the knowledge that exists throw up a good many questions and dilemmas. Supported recently by empirical evidence (Chemi 2014, Holst 2015), this awareness has been voiced in the theoretical texts that have been most influential within teacher and pedagogical colleges: Austring and Sørensen (2006), and Hohr and Pedersen (2001).

In the Nordic region one direct inspiration for the idea of cultural partnerships is to be found in the Norwegian school's *Cultural Rucksack*. Borgen and Brandt (2006) and Borgen (2011) have evaluated the Norwegian initiative and have shown that the project was characterised by high administrative complexity – something not found in the Culture Laboratory project.

Moreover, the Norwegian studies were based on quantitative methodologies, having chosen a mapping method based on questionnaire surveys, whereas the present Danish research collected qualitative, narrative and polyphonic data.

Looking at Scandinavian cases is significant for a broader international perspective, as these countries rest on a long-standing pedagogical tradition that values democratic formation and *Bildung*. In the context of the Nordic models of welfare, pedagogical examples that are attentive to sociality, expression and hands-on experiences have flourished. Despite that, arts-integrated programmes and artist-school partnerships are still a challenge. This gap makes the present study perhaps interesting for a larger community of arts-based practitioners, as it can bring empirical descriptions of successes and challenges in these practices.

3.5 Methodology

Based on the above-mentioned national requirements and international trends, the Culture Laboratory research project sought to achieve a sharper focus on and insight into the following research issues:

- What characterises the encounter between children and arts and culture in schools when professional artists are involved?
    - What kinds of activities are offered?
    - What kind of art and expression?
    - Which collaborators?
    - What learning outcomes?
- How are arts and culture integrated into schools?
    - What challenges does this encounter entail in the school's organisation, activity planning and execution?
    - What reflections do teachers make when arts and culture are integrated into schools?
    - What considerations do artists and cultural institutions make when they cooperate with schools?
    - For what purpose are arts and culture involved in schools?
- What does it mean for students' learning and development when schools integrate arts and culture into their activities, according to teachers, artists and cultural institutions?

- What skills are required in the children, in the artists, in the cultural institutions/cultural mediators and in the teachers within partnerships? — and what skills does the encounter develop in the children and in the adults (teachers, artists, cultural mediators)?

These were the guiding research questions for the project, upon which the empirical study was designed. For research into the subject field, described as artist-school-and-culture partnerships, a dual-track approach was proposed, in which both qualitative collection of knowledge about these partnerships, and collaborative approaches (action research-inspired) were used. Research aimed at hearing all the voices in the encounter (students, school staff and the artists or cultural institutions working with the schools), in order to draw the full picture in this area. In interviews and field observations I was looking at differences and similarities. In other words, throughout the project, the aim was to systematically, thoroughly and scientifically describe the phenomenon of artist-school-and-culture partnerships, to subsequently analyse and interpret the material, thereby highlighting strengths and challenges.

In order to describe the research, methodological considerations and actions, it is important to look at the overall structure of the project. The following units covered the project's work areas:

- project management (facilitation of the specific activities in the subprojects and overall coordination),
- development (design and execution of the subprojects) and
- research (documentation, analysis and reporting).

Three independent entities, responsible for each area, maintained a close and on-going dialogue, but remained autonomous in their decisions. This organisation ensured a high level of independence and positive cooperation. The three areas/units can be visualised as Figure 3.1.

The project management unit with project manager Cecilia de Jong was located at the Culture Region Funen. Its role was central to initiating the project, and taking the fundamental responsibility for coordinating the concrete development activities. This coordination included the recruitment of participants, the matching of schools with artists and cultural institutions, the arrangement of several large-scale events (kick-off development-workshop, final conference, and study group), maintenance of informal conversations and knowledge sharing on social media, bridge building between research unit, development unit and Culture Region.

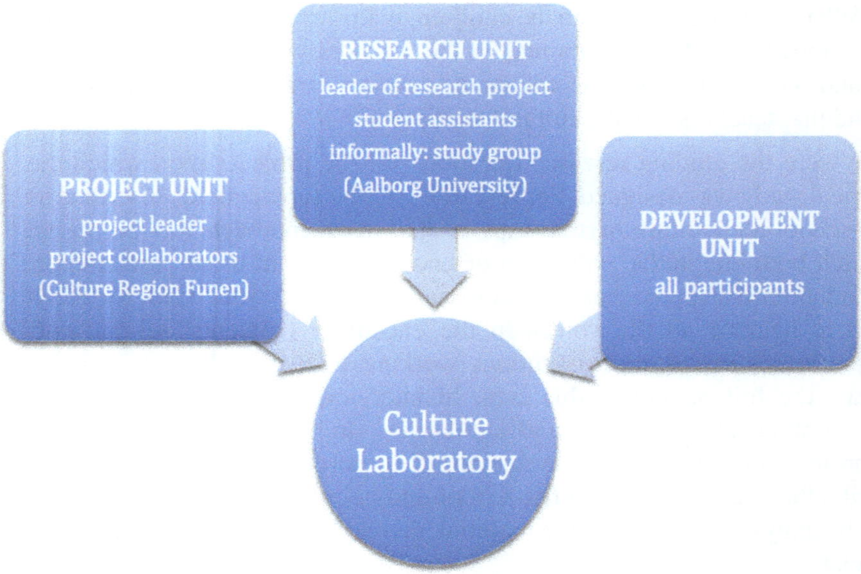

Figure 3.1 The organisation behind Culture Laboratory.

The research unit was located at Aalborg University, at the Department of Learning and Philosophy, with myself as a research project manager. It was the responsibility of the research unit to design a plan for research (research design), carry out documentation according to the plan (empirical collection) and disseminate the results (final report in Chemi 2017b, present contribution and final conference). The research unit had collaborated with the project unit before the project was granted funding for development and research. This early collaboration was focused on the conceptualisation of the content framework for the project, based on existing research findings and knowledge. This on-going dialogue took place informally but regularly between October 2014 and December 2015. The formal collaboration between the Culture Laboratory, the Culture Region and Aalborg University entered into force on January 1, 2016 and ended on June 30, 2017. The collection of empirical data at the nine selected schools and cultural institutions took place between April 2016 and December 2016. As a researcher, I had full autonomy in my decisions on research design. In other words, I made all the final decisions about what was interesting to investigate and how it would be appropriate to investigate it. The Culture Region's council approved the research design in advance.

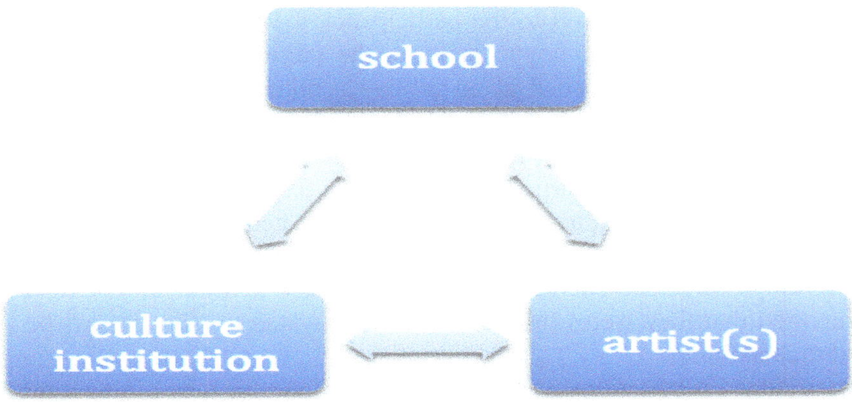

Figure 3.2 The partnership.

The development unit was located at the participating parties: schools, artists and cultural institutions. Each participant took part in the project through a representative: a teacher, an artist, a cultural facilitator or mediator. The schools contributed by usually involving a single class of about 20 students and only in two cases by involving two reciprocally collaborating classes. Cultural institutions in most cases involved the employee who was responsible for teaching, here defined as cultural facilitator or mediator. Each partnership consisted of a trio that can be visualised as in Figure 3.2.

The participants were selected on the basis of a number of different criteria: their will and commitment to the activities, geographical variation across the region, topographic variation (city, village), school size, classroom variation (junior, senior or middle stage), variation of art and genre, institution type (museum of cultural history, local history or art museum, library, symphony orchestra, theatre). In addition, the project unit, on the basis of logistical, contextual and intuitive criteria, sampled the individual participants. It should be emphasised that the logistic criteria did not exclude opportunities for participation in the activities. In some cases, the activities required the class to use transportation to a given cultural institution or site; the project unit covered this expense, when necessary. The composition of the different partnerships became a kind of (happily) arranged marriage: an arranged partnership.

The responsibility of the development unit was to design and carry out the activities that the students should participate in. Because the selection and planning of the activities, according to the project description and the initial meetings, was to be left to the participants, the project and research units facilitated an initial dialogue between the participants. On April 5, 2016, the project was launched at an all-day seminar consisting of academic presentations on aesthetic learning processes, as well as facilitated and informal conversations between the partners. The facilitated conversations were intended to allow all participants to discuss and negotiate their joint activities. As part of the activities, this meeting created the organisational framework for free conversation to occur in the partnerships. The bottom-up approach empowered and energised the participants, who seemed to be (and later reported that they were) fully ready and motivated to start the artistic and cultural activities.

The arranged partnerships were not revealed before this kick-off meeting and the three participants were given the opportunity to get to know each other and plan an initial outline of activities. As the interviews show, these conversations were fundamental to the further development of subprojects, to the extent that all partnerships reached binding agreements on the content of the subproject. In some cases, the agreements were concluded with specific determination of content and choice of activities. In the interviews, the participants reported that their conversations focused on clarifying the following:

- What are my skills (What can I do? What are my favourite artistic/cultural areas)?
- What are my interests (What do I want? What is my benefit?)
- What are my needs (concrete framework for cooperation)?
- What are my visions for the joint project (what can we do together)?

In addition, a more collaborative conversation was initiated focusing on the following:

- What opportunities can we see in our cooperation so that our different skills and interests can come together?

These open conversations ended with concrete agreements, which were structured differently, but all of which addressed:

- When the participants should continue the conversation
- How they should communicate in future
- Possible dates for subproject activities
- Outline of activities and their content.

In order to fulfil the research objective, qualitative and collaborative methods were chosen as working methods and tools. The research was limited to primary school children (6–16 years) and to the nine selected artists and cultural institutions that worked with the target group. As mentioned above, partnerships were selected on the basis of qualitative criteria for diversity and representativeness, including, for example, geographical spread (city/village, different regions and regions), age range (different stages), diversity in learning offers (different types of experiences), and the number of teachers/pedagogues and children.

Qualitative data were collected by means of ethnography-inspired classroom observations, semi-structured interviews (students, artists, cultural institutions) that were recorded and transcribed, action research-inspired study groups (recorded and transcribed), and document and artefact analysis.

3.6 Findings

The purpose of this research has been to develop both research and practice, as it was aimed at researchers as well as practitioners. The results of this project thus cover both research documentation (development of research criteria and categories to evaluate similar projects) and the development of knowledge for practice (qualification of practice). Although the research and evaluation results cannot be clearly separated, the results of the project can be separated into a research level (research knowledge) and a level of development (practical knowledge).

These findings are placed at the intersection between research-based evaluation (research and theory development) and evaluation-based research (documentation and practice knowledge). First of all, I developed research categories in order to investigate this complex field (artist-school-and-culture partnerships) and the two-fold phenomenon of art and learning. It is always a difficult task to evaluate and document the benefits of interdisciplinary initiatives and complex learning environments. The researcher's work often ends up being a reduction of this complexity by identifying specific research categories and placing the participants' statements within these analytical categories. The following categories of values provide a research basis for formulating qualified interpretations of empirical data and for qualifying the partnerships' practice knowledge. I identified the following categories in relation to the four levels of student, teacher, cultural mediator, and artist:

Student perspective

Positive emotional response
Learning outcomes

Positive emotions
Negative emotions
Otherness
The specifics of the arts and aesthetics

Teacher perspective

Teachers' perception of children's benefits (emotions)
Learning outcomes (cognition)
Teachers' own perspective
Partnership and cooperation with artists

Cultural mediator perspective

What worked well
Learning outcomes for students
Learning outcomes for cultural institutions
Educational design
What is special for cultural institutions

Artist perspective

Output for students
Output for artist
Educational design
What worked well
How artists think

These research categories have assisted in harvesting a number of findings that show what specific learning outputs all the participants experienced. To sum up all of them would go far beyond the scope of the present contribution. Instead, I wish to focus on a specific set of findings, which demonstrate a particular emotional response in the students' experience: emotional arousal. This response is described as surprising, exciting, novel and different, and brings with it a number of learning outputs.

3.7 "This Is Really Cool"

A number of elements are mentioned as positively experienced in the students' interviews. From generic to very specific, these positive responses recurrently emphasise the student's emotional *arousal* in a sudden occurrence that is regarded as new, surprising and exciting.

As Bamford (2006) conceptualised when she drafted the UNESCO report on children's experiences with artistic and cultural activities, a common feature of these encounters with the arts was the 'wow' effect (*the wow factor*), i.e. an unexpected, positive and surprising experiential recognition. The students interviewed in the present study also reported that they were surprised by some activities in their artistic expression as such. For instance, the smoke sculptures exploded by the sea by a conceptual and land-artist provoked great excitement in the schoolchildren. In this case, the activity is just designed to surprise the recipient, who merely participates as spectator. However, the students also reported surprise and excitement resulting from their own works of art, in which they participated as makers. A 2nd grade boy explained his positive experience, reporting his amazement about the process and the product of the artistic activity: "I think it was cool because everything went up in smoke there, [the smoke] could have different colours ... and then the picture we made working together, I think it was really nice."[2] As Berlyne (1971) conceptualised, *arousal* is a psycho-emotional experience or state that is fundamental to learning, because it brings with it a heightened level of attention, and perhaps also of concentration. Berlyne's arousal theory recognised a special role for emotions in learning processes. An individual that is aroused by a specific experience is able to perceive alertness and focused attention, a state that may facilitate optimal absorption of learning matter. According to Ellen Winner (1982), arousal from participation in the arts has a double effect, on the one hand, hedonistic, because of the "pleasure given by art's formal properties" (Winner 1982, p. 64), on the other, cognitive: "Art serves the human need for knowledge [and] functions ultimately to reveal and clarify reality" (Winner 1982, p. 65). Similar arguments are echoed in Vygotsky (1997), where learning is seen as a conscious and qualified response to stimuli, appropriate to a given context. Vygotsky defines this process as the emergence of purposive behaviour and it is central to explaining how emotional arousal can generate learning outputs. If learning is understood as embodied and emotional, then psycho-emotional awakeness would provide fertile ground for any learning process. "The perceptive system, based on sensory and bodily experiences, feeds in different ways the cognitive system, for instance by triggering arousal or interest or engagement. In this way organisms apply analytical skills, initiating processes of decomposition and composition, but also appropriate responses in form of

[2]From now on, all quotes without reference will refer to the original interviews in this project.

actions" (Chemi 2017a, p. 19). As Lynn Fels (2011) notes, the concept of "wide-awakeness" (p. 340), already present in Maxine Greene's pedagogical philosophy, is fundamental to artistic (and specifically performative) inquiry.

These experiences seem, by means of surprise, to awaken the students because they are designed to provoke awe in themselves, for instance using sudden explosions, changing colours, displacements of artefacts. They also surprise the students because they are experiences that are new to them. Novelty value is indeed one of the elements that has most positive appeal to children. Sometimes novelty is appreciated for itself, regardless of content. In other words, children might appreciate positively a change in school routine, regardless of what it is about. In this case, the arts have not much to contribute, except their inherent capacity of bringing novelty to the surface or working with continuously new or renewed approaches. When students mention the usefulness of new experiences, they specify that, for them, it is fun to try something either in production (2nd grade boy: "We had not tried it before") or in a new physical environment (3rd grade boy: "Fun to see something else"), or just something new in relation to school and school tasks, as a 3rd grade boy says: "It has helped a lot that we didn't just sit and do the same things all the time, but did something new and saw it all from several angles". A 9th grade girl reported a similar experience: "I think it was fun. It was challenging and it was fun trying something different, something that we are not used to [in] everyday life". Several students emphasised how exciting it was to try new media (8th grade boy: "It was exciting, we have been around and learning a lot of new stuff about stop-motion and generally on how to work with animation. Also very funny when we listened") or maybe other skills (a 9th grade girl: "[It was] great fun to focus on something different from what we are just used to in school. It involves a lot of school-subjects, such as mathematics and Danish, [but here] there is also a focus on personal communication").

Much of the students' enthusiasm was justified by their impression of the artistic and cultural activities as different, or *other*, from school. The students' formulations ranged from a generic "change in the school day" (9th grade boy) to a more nuanced explanation of how the activities experienced differed from the everyday life of the school: "[It was] quite fun to go somewhere other than school and learn something different from what you can do in school" (2nd grade girl), "You don't have to sit on a chair *aaaaall* the time and you could play a little" (2nd grade boy) "So it's a bit more fun than just sitting on your chair, so it's a bit more fun when you get out and do something else" (7th grade girl). In other words, the students' arousal seems to emerge from the fact that the artistic experiences and the approaches used in Culture Laboratory were *new* to them and also *unexpected* in a school context.

3.8 Different from School

Whether the students assessed their experience positively or not, they all took a clear position on the extent to which the activities of the Culture Laboratory differed from their everyday life at school. All students were asked explicitly to consider the following:

What is the difference between ordinary art education and these partnerships for you?

What is the difference between the regular school and these activities for you?

The interviews showed a fairly clear awareness that there was actually a difference and which specific elements defined this differentiation. The students' answers on this specific topic can be summed up according to the following categories:

- Different from schools
- Just something new
- Different use of technology
- No time pressure
- Hands-on
- Different from visual art classes in school.

Different from schools – The students described their school day as characterised by tasks, routines and repetitions. This was considered less engaging than the activities they experienced in the art and culture project. With a clear tone of criticism, a girl said: "We are used to, like ... sitting and doing the same thing ... otherwise, when we are in groups, we are often outside and do something where we can run ... and we must shout to each other ... But what we did today, that's in every way ... different" (2nd grade girl). In other cases students considered the presence of repetition in artistic and cultural activities negative and demotivating (they mention "boredom"), where this student says repetitions belong to school life.

When I asked the girl to explain what she meant by "sitting and doing the same thing" in school versus the repetitions that accompany artistic making, she responded as follows: "In one way we did not do the same because we were many different groups who did something different. And so, there was someone with whom to share". A surprising point came from a 2nd grade boy who perceived activities as quieter than schoolwork, despite the activities being designed as group work and as active, he said: "I think it was a bit quieter than it normally is". Other students thought the difference was that the children could walk about a little more and there were not many children other than those they already knew.

Just something new. – Most often, the children's statements were merely indicators of a generic difference, for example, a 2nd grade girl said: "[normally] you just have to sit inside a warm room all day. There, we were just outside all day. It was something new. It was really nice to try something new too". Other times, children wanted an opportunity to learn subjects other than school subjects: "it was fun learning something new, rather than just maths and Danish and such" (2nd grade girl); "They are not boring hours, like the ones where we have to write and read and do maths" (3rd grade girl). The principle of variation was mentioned by a boy in the 8th grade: "It may also seem difficult to sit on it, on the chair, all day long and just keep getting information into our heads. It's also because we have these Tuesdays [special activity-days] where we go out, for example, and then we relax ... or do not relax, but like that, we get away from everything, from the sitting still".

Different use of technology. – Another point that differentiated schooldays from the arts projects was the latter's approach to technology. In the students' interviews, it was noted and appreciated that a different approach to digital media was cultivated in these artistic activities. This seems to be different perhaps because of a more applied method and because this application is more related to reality. A 2nd grade girl said: "we could find out where we could use something from some stories, and then we got some themes for the [performances at the] theatre so we could figure out what it should be".

No time pressure. – Several students mentioned experiencing the overall atmosphere as more relaxed in these projects than at school, and believed that it was due to the lack of time pressure in the projects. A 2nd grade girl said: "yes, it's such a time-out, somehow" and a 3rd grade boy reported about the positive repetition of the same school tasks, but throughout a whole day: "we only have this subject ... here we do only that the whole day".

Hands-on. – An 8th grade boy talked about what several students seemed to notice as a difference between school and the artistic and cultural projects:

> What you learn here is something that you could also learn in a subject called *history*, but what you learn here is something you do

> not learn at school, because it's something completely different. Something quite different from what it is in a school because school, it's more about books and computers. Here, you can learn something while doing something, like an active learning. And it's the same too when you look at trains and read a bit about them. Also that's where we did stop-motion activities ... It's not just looking at something and then pressing a screen and then sitting for almost three hours and then looking ... We also began to [learn how to] make movement, started doing some drawings, making something, putting some figures together, and something like that, and it became a little fun, it became easy to learn ...

According to this quote, the students learned *while* doing something practical, thereby learning to trust that the process unfolds from an immediately meaningless repetition to a process that starts to make sense. The activities are the focus of the projects and are not school assignments. Another version of this statement is the following: "you almost do not learn anything, so ... no homework ... Then you leave it and you have fun too" (3rd grade girl). It is interesting to note what the girl says: "you almost do not learn anything" in the projects. This indicates that the students might have a very limited view of what learning is when it takes place at school, e.g. limiting it to homework, tests and tasks.

Different from visual art classes in school. – Surprisingly, the children also pointed out that the projects experienced were far from what they encountered in visual arts at school. These statements suggest that the learning outcomes our research was able to document were not solely due to the artistic and cultural activities in themselves, but to a particular design of them. In other words, what ignited the students' engagement might not be the arts, but a specific approach to the arts. According to the students, in school, visual arts education appears to be characterised by boredom and unoriginality, as this exchange among three 2nd grade students shows:

> Boy 1: We usually only sit and draw ...
>
> Boy 2: ... and draw lines.
>
> Girl: ... and then we sit with this book. We usually spend a minute and then we draw ... We also tried to draw with pencil on paper, so we did not see what we were drawing. We always do that when we start. But otherwise you can decide what you want to do ...

A third grade boy elaborated on this perception by pointing out the possibility of deep immersion in the art project he experienced: "[it is] different, also because you get more into it and you get to know more". Students seemed to relate positively to the challenges and tests of new artistic media and techniques, and maybe even missed the same opportunity in their normal school-day, as the exchange among these 3rd grade students shows:

> Boy 1: When we have visual arts with [our teacher], we don't do so many paintings, so it's very much like ... drawings where something happens...
>
> Girl: Yes, I thought so too.
>
> Boy 2: It was much more fun to do that with [the artist].

Unfortunately, it does not seem that visual arts education – as the only art form spontaneously mentioned in student interviews – is something that cognitively challenges or emotionally fascinates children. On the contrary, it appears to be negatively assessed in relation to the artistic and cultural activities of Culture Laboratory. In other words, when students compare the "normal" artistic activities in school and their experiences with the arts within Culture Laboratory, the comparison is all to the school subject's disadvantage. Possibly, this assessment suggests a didactic and pedagogical under-prioritisation of artistic subjects in school. As a 9th grade girl candidly says, visual art is a minor activity, done when the more important subjects are covered: "[We did art at school] we once did it when there was nothing else we could do".

3.9 Broader and Future Perspectives

The broader results of the project can be summarised as follows: the activities in the artistic and cultural partnerships are perceived as positive and engaging by all participants involved, but for various reasons. The children's positive experience was mainly due to the novelty in their school day, as a breach of school routine and didactic approach. This surprise (*wow factor*) depends essentially on how frequently the students are accustomed to artistic and cultural activities, and therefore on how new and surprising *to them* is the encounter with the arts, and secondly on how many of these activities differ qualitatively and methodologically from the school day. Qualitatively, the artistic activities seem to offer (better) opportunities for social relationships in playful settings and team/project work, and they are more hands-on and experience-based practices, as well as actively involving the students.

The teachers' positive assessment of the project's activities is mainly due to their observation of the students' benefit and positive experience. The teachers' own experience is closely linked to and dependent on the learning outcomes of the children and the usefulness of the activities in a school context: they are happy when the students seem to learn educational content, gaining acknowledgments and inspiration from the external participants, and when they seem to apply appropriately the knowledge they receive. Teachers are mostly concerned about logistical issues, such as the prevention of accidents, and children's safety, good discipline, and time frames, but also about a few conceptual elements, such as the activities' cross-disciplinary content, the dialectic product/process, and the freedom of participation for children.

The positive assessment of the cultural institutions is mainly due to a logistical agreement within the partnership and the great commitment to the partnership. Participants from these institutions mention well-developed logistics and laid-back negotiations in order to agree on the activities' frames. Last but not least, they mention the high energy sensed in the reciprocal collaboration.

The artists' positive assessment is due to a variety of qualities that span widely around the partnerships. This is not surprising, as artists tend to be sensitive to the observation of qualities and to the attention to multiplicity (Chemi, Jensen & Hersted 2015) in their daily work. The artists here interviewed used several metaphors to express this diversity of forms, such as an "opportunity room" or a "conceptual umbrella" or a place for "possibilities". They also pointed to the content of the partnership as conversation-based, interdisciplinary, narrative, co-creative, and professional.

Even though the artistic and cultural partnerships seem to have generated learning across all participants, this happened with a large diversity of experiences. The participants emphasised and valued very different learning outputs. To specifically develop this point would lead us too far from the main concern of the present contribution. However, these dissimilarities call for further reflection on the projects' influence on the participants' learning and development, especially if these projects are to be repeated in the future. As a matter of fact, as a consequence of this experience, the Funen Culture Region will initiate 36 new partnerships in the school-year August 2017–June 2018. Hopefully, the new partnerships will make use of the lessons learned from the present project and will address the clear benefits as well as the – equally clear-downsides.

All the participants expressed what was particularly perplexing in the projects' activities. Their replies show the need for further development at the practical level of educational and artistic design of the activities, but also the need at theoretical level to further investigate the collective learning that might emerge. At practical level, the challenges mentioned spanned from the need for time to set up and clean up art materials or to cope with the repetitions that were required in the artistic processes, to the understanding of new tasks, of idea generation and problem solving (students), or from the necessary considerations about timing (the activities should not be placed late in the day when the children are tired or should not collide with other seasonal activities, such as Christmas celebrations), to security for the children's physical and psychological well-being – which may also limit authentic artistic experiences (teachers).

Participants from cultural institutions were, on the one hand, concerned about over-doing things (activities that are too ambitious, not concrete enough or lack interactivity) and, on the other, about under-stimulating the students, for instance, allowing inactivity by some students or accepting unprepared teachers. These participants were aware that experiences in and with cultural institutions can be so new for students that they end up provoking anxiety or distracting from learning activities. Paradoxically, in these cases, the arts and culture might reduce or weaken the intended learning outputs. Once again, this topic is far too complex to fully address here, because it is contextual to learning values and practices that do not necessarily match traditionally designed school results. According to artists, what is critical in these partnerships is the careful design of activities that must balance between a professional approach to artistic activities and room for experimentation. In this way, one should aim at avoiding strict 'recipes', without giving in to frames that are too free and improvisational.

All participants were finally asked to formulate specific recommendations for addressing the challenges above. Most surprisingly, even young students could be very detailed in their formulations. What they suggested to hold on to was the positively-felt organisation of activities that allowed for self-determination and self-organisation, together with the wide opportunity for receiving help and support from several parties (from home, teachers, artists, peers).

All participants suggested carefully considering time frames that allowed for sufficient preparation and development of the activities. Several maintained that focus should be on the quality of artistic experiences and ambitious purposes (philosophical or foundational purpose, usability or applicability).

Last but not least, participants indicated that the following conditions were fundamental to the positive results of the activities and strongly suggested further developing them in future partnerships: clarification in advance (as artists mention, it is important to answer to intentionality: "what do we want?"), possibility of sharing materials and knowledge, project benefits should be systematically documented before (project description) and after (evaluation).

In conclusion, I wish to suggest that experiences and knowledge from the Culture Laboratory, or broadly from the Open School partnerships, might contribute to reframing the role of the arts in formal and informal learning environments and point to possible directions for future developmental practices and theoretical investigations.

References

Austring, B. D., and Sørensen, M. (2006). *Æstetik og Læring: Grundbog om Æstetiske Læreprocesser*. København: Hans Reitzels Forlag.

Bamford, A. (2006). *The Wow Factor: Global Research Compendium on the Impact of the Arts in Education*. New York, NY: Waxmann.

Bamford, A., and Qvortrup, M. (2006). *The Ildsjæl in the Classroom: A Review of Danish Arts Education in the Folkeskole*. Aakirkeby: Kunstrådet.

Berlyne, D. (1971). *Aesthetics and Psychobiology*. New York, NY: Appelton-Century-Crofts.

Borgen, J. S. (2011). "The cultural rucksack in Norway: does the national model entail a programme for educational change?," in *The Routledge International Handbook of Creative Learning* eds J. Sefton-Green, P. Thomson, K. Jones and L. Bresler (London: Routledge), 374–382.

Borgen, J. S., and Brandt, S. S. (2006). *Ekstraordinært Eller Selvfølgelig? Evaluering av Den Kulturelle Skolesekken i Grunnskolen.* Oslo: NIFU STEP.

Bresler, L. (ed.). (2007). *International Handbook of Research in Arts Education*. Dordrecht: Springer.

Chemi, T. (2014). *The Art of Arts Integration: Theoretical Perspectives and Practical Guidelines*. Aalborg: Aalborg Universitetsforlag.

Chemi, T. (2015). *Den Åbne Skole Mellem Læring og Kunst. Folkeskolen*. Available at: https://www.folkeskolen.dk/578038/den-aabne-skole-mellem-laering-og-kunst-partnerskab-konflikt-eller-hoeflig-samtale-

Chemi, T. (2017a). "A safe haven for emotional experiences," in *Innovative Pedagogy: A Recognition of Emotions and Creativity in Education* eds T. Chemi, S. Grams Davy and B. Lund (Rotterdam: Sense Publishers), 9–25.

Chemi, T. (2017b). *Partnerskaber Blandt Kunstnere, Kulturinstitutioner og Skoler: Kulturens Laboratorium*. Aalborg: Aalborg Universitetsforlag.

Chemi, T., Jensen, J., and Hersted, L. (2015). *Behind the Scenes of Artistic Creativity. Creating, Learning and Organising*. New York, NY: Peter Lange.

Csikszentmihalyi, M. (1990). *Flow: The Psychology of Optimal Experience*. New York, NY: HarperCollins.

Csikszentmihalyi, M. (1996). *Creativity: Flow and the Psychology of Discovery and Invention*. London: HarperCollins.

Davies, D., Jindal-Snape, D., Collier, C., Digby, R., Hay, P., and Howe, A. (2013). Creative learning environments in education—A systematic literature review. *Think. Skills Creat.* 8, 80–91.

Fels, L. (2011). "A dead man's sweater: performative inquiry embodied and recognized," in *Key Concepts in Theatre/Drama Education,* ed. S. Schonmann (Rotterdam: Sense Publishers), 339–343.

Fiske, E. B. (1999). *Champions of Change: The Impact of the Arts on Learning*. Available at: http://artsedge.kennedy-center.org/champions/pdfs/ChampsReport.pdf [accessed 21, 2017].

Fleming, M., Bresler, L., and O'Toole, J. (eds). (2015). *The Routledge International Handbook of the Arts and Education*. London: Routledge.

Gardner, H. (1993). *Creating Minds: an Anatomy of Creativity Seen Through the Lives of Freud, Einstein, Picasso, Stravinsky, Eliot, Graham and Gandhi.* New York, NY: Basic Books.

Gardner, H. (1994a). *The Arts and Human Development: a Psychological Study of the Artistic Process.* New York, NY: Basic Books.

Gardner, H. (1994b). *Frames of Mind: the Theory of Multiple Intelligences.* London: Harper Collins.

Goodman, N. (1968). *Languages of Art: An Approach to a Theory of Symbols.* Cambridge: Hackett Publishing.

Hohr, H., and Pedersen, K. (2001). *Perspektiver på Æstetiske Læreprocesser (Perspectives on Aesthetic Learning Processes).* Frederiksberg: Dansklærerforeningens Forlag.

Holst, F. (2015). *Kortlægning af Forskning i Effekten af Børns og Unges Møde Med Kunsten*. København: Statens Kunstfond.

KL (2015). *Læring i Den Åbne Skole*. Oslo: Kommuneforlaget.

Marshall, J., and Donahue, D. M. (2014). *Arts-Centered Learning across the Curriculum: Integrating Contemporary Art in the Secondary School Classroom*. New York, NY: Teachers College Press.

Ministry of Culture (Kulturministeriet) (2014). *Strategi for små Børns Møde med Kunst og Kultur*. Available at: http://kum.dk/uploads/tx_templavoila/kum_brochure_smaborn_klausuleret%20(3).pdf [accessed April 6, 2014].

Ministry of Culture (Kulturministeriet) (2017). *Kulturregioner*. Available at: http://slks.dk/kommuner-plan-arkitektur/kulturaftaler/kulturregioner/ [accessed April 6, 2014].

Ministry of Education (Undervisningsministeriet) (2017). *Folkeskolens Formål*. Available at: http://www.uvm.dk/folkeskolen/folkeskolens-maal-love-og-regler/om-folkeskolen-og-folkeskolens-formaal/folkeskolens-formaal [accessed April 6, 2014].

Nationalt Netværk af Skoletjenester (2015). *Skolens Transport af Elever til Kulturinstitutioner og Eksterne Læringsmiljøer: En Kortlægning af Nationale Erfaringer fra Skoler, Kommuner, Kultutinstitutioner og Andre Eksterne Læringsmiljøer*. Available at: http://skoletjenestenetvaerk.dk/wp-content/uploads/2016/01/Skolers-transport-af-elever-PIXI.pdf [accessed April 6, 2014].

Nationalt Netværk af Skoletjenester (2016a). *Samarbejder og Partnerskaber Mellem Kommuner, Kulturinstitutioner og Eksterne Læringsmiljøer: En Kortlægning af Nationale Erfaringer fra Skoler, Kommuner, Kultutinstitutioner og Andre Eksterne Læringsmiljøer*. København: Nationalt Netværk af Skoletjenester.

Nationalt Netværk af Skoletjenester (2016b). *Skolers Brug af Undervisningstilbud på Kulturinstitutioner/Eksterne Læringsmiljøer: En Kortlægning af Nationale Erfaringer fra Skoler, Kommuner, Kultutinstitutioner og Andre Eksterne Læringsmiljøer*. København: Nationalt Netværk af Skoletjenester.

Schonmann, S. (ed.). (2015). *International Yearbook for Research in Arts Education 3/2015: The Wisdom of the Many-Key Issues in Arts Education*. Münster: Waxmann Verlag.

Sefton-Green, J., Thomson, P., Jones, K., and Bresler, L. (eds). (2011). *The Routledge International Handbook of Creative Learning*. London: Routledge.

Sæbø, A. B. (ed.). (2016). *International Yearbook for Research in Arts Education 4/2016: At the Crossroads of Arts and Cultural Education: Queries Meet Assumptions*. Münster: Waxmann Verlag.

UNICEF (1989). *Convention on the Rights of the Child: Adopted and Opened for Signature, Ratification and Accession by General Assembly Resolution 44/25 of 20 November 1989 Entry into force 2 September 1990, in Accordance with Article 49*. Available at: http://www.ohchr.org/en/professionalinterest/pages/crc.aspx [accessed April 6, 2014].

Vygotsky, L. S. (1997). *Educational Psychology*. Boca Raton, FL: St. Lucie Press.

Winner, E. (1982). *Invented Worlds. The Psychology of the Arts*. Cambridge: Harvard University Press.

Winner, E., Goldstein, T. R., and Vincent-Lancrin, S. (2013). *Art for Art's Sake? The Impact of Arts Education*. Paris: OECD.

4

Designing Activities for Teaching Music Improvisation in Preschools – Evaluating Outcomes and Tools

Una MacGlone

Reid School of Music, Alison House, 12 Nicolson Square, University of Edinburgh, Edinburgh, EH8 9DF, Scotland

Abstract

This chapter gives an in-depth analysis, through Activity Theory, of two key music improvisation activities developed by the author. These two activities were part of a series of music improvisation workshops delivered by the author to a group of six preschool children in Scotland. The workshops were designed around two novel constructs back-engineered from the researcher's professional experience as an improvising musician, Creative Musical Agency and Socio-Musical Aptitude. Creative Musical Agency (CMA) is: The child creates novel musical material independently and executes this in the group improvisation. Socio-Musical Aptitude (S-MA) is: The child creates a musical response with reference to another child's musical idea in the group improvisation.

Through examining video analysis of the workshops, teacher interviews, children's talk data and the author's own reflexive data, a rich picture of the workshop activities is gained. The theoretical lens of Activity Theory revealed the creative musical decisions the children made and the ways in which these were mediated through physical and symbolic tools. Interesting possibilities and challenges in the activities were explored, and, therefore utilising Activity Theory has great potential for other researchers to examine complex creative pedagogical contexts.

4.1 Introduction

Music improvisation pedagogy is a rapidly developing field, which, can be seen to reflect a music profession where musicians are increasingly expected to be able to improvise (Johansen, Larsson, MacGlone, Siljamäki, 2017). In this chapter, I will introduce two novel theoretical constructs, which I designed for the purpose of teaching music improvisation in preschool settings in Scotland. These constructs, Creative Musical Agency (CMA) and Socio-Musical Aptitude (S-MA) are developed from my professional experience as an improvising musician. The empirical work described and discussed in this chapter is from my PhD, in which I designed, delivered and assessed two six-week programmes of improvisation workshops for preschool children in an Action Research framework. This chapter will focus on my critical reflection on two key workshop activities through the theoretical-analytical lens of Activity Theory and discuss the implications for music improvisation pedagogy with preschool children.

4.2 Previous Literature

4.2.1 Understandings and Applications of Improvisation in Music

An important issue in researching musical improvisation and any of the contexts in which it appears, is apprehending the diversity in understanding of what constitutes improvisation. Improvisation could be understood as virtuosic extemporizing as played by saxophonist Evan Parker, a clinical process in Music Therapy, a teacher encouraging group creativity in a classroom as generative process in composing, or, a parent trying various invented melodies or sounds to help their baby sleep. These represent various contexts of improvisation: artistic, therapeutic, pedagogical, and, everyday. All of these different applications of improvisation can be appreciated as the negotiating or manipulation of unanticipated events, corresponding to the Latin roots of the words itself: *improvisus* – unforeseen. However, in music, the extent to which the events are completely unknown varies greatly depending on genre and context. Improvisation in music is found in a diverse range of genres, for example, church organ music; jazz; contemporary classical; Indian classical music etc. and creative contexts, yet offers a distinct function in each situation (MacDonald, Wilson & Miell, 2012).

The artistic practice best reflecting my professional background as an improviser is known as free improvisation, understood to be a distinct practice which prioritises the socio-musical aspects of music making (Lewis, 2000), where creativity is situated largely with the performer rather than composer or score (Wilson & MacDonald, 2016). Free improvisation has been characterised as enigmatic by some authors (Ashley, 2009). This concept aligns with often quoted musician/writers Braxton, (1985) who highlighted the ubiquitous practice yet under-acknowledged aspect of improvisation and Bailey, (1992) who mostly avoided definitions, instead describing a variety of improvisation contexts (Lewis & Piekut, 2016). These contrast with conceptualisations of improvisation as an everyday human process (MacDonald et al., 2012 & Lewis & Piekut, 2016).

As we see, there is no commonly used definition (MacDonald et al. 2012), and writing on improvisation has only recently (since within the last 25 years) been a subject of scholarly interest, usually published by small, hard to find companies (Rose, 2012). A reason for the many definitions could reflect on different gatekeepers who have conflicting views and agendas (Johansen et al. 2017). Increasingly, contemporary writing on free improvisation has begun to either search for common features across context, such as creativity and spontaneity (Hickey, 2015) or to highlight key aspects which are distinct to the context of the research, for example, interactive aspects (Linson, 2014).

4.2.2 Existing Approaches in Improvisation Pedagogy

Increasing numbers of researchers, practitioners and institutions ranging from orchestras to educational establishments are interested in improvisation as a creative practice and developing methods and approaches to teaching improvisation (MacGlone & MacDonald, 2017; Heble & Laver, 2016; Lewis, 2000). This could be a result of new curricula (within in the last ten to fifteen years) which emphasise creativity, student centred teaching and process based learning, for example in Scotland (Education Scotland, 2006). These curricula have been designed with the aim of preparing children to participate fully in changing world (Education Scotland, 2006) and encourage 21st century skills. These skills can include "creativity, critical thinking and problem solving, collaborative skills, information technology skills, and new forms of literacy, and social, cultural, and metacognitive awareness" (Griffin & Care, 2014 p. 14).

When comparing different approaches to teaching improvisation, complexity in the heterogeneity of participants and intended outcomes presents a challenge in building overarching theories. A variety of factors such as age of children/students, educational approach and research methodologies all contribute to this complexity. Therefore, through examining key texts with the aim of exploring the reported pedagogical approach is an effective way of understanding these diverse contexts. Within pedagogical applications of improvisation, teachers' and researchers' own beliefs or internalised cultural or genre based beliefs about improvisation may affect how and what they teach, a common belief being that students have to reach a certain level of technical skill before they can improvise (Whitcomb, 2013; Koutsoupidou, 2008). A contrasting view is that children are "natural" improvisers (Hickey, 2009; Barrett, 2006; Young, 2003). These beliefs relate to the particular orientation of the pedagogical approach, examination of the literature concerned with teaching improvisation revealed two broad categories of pedagogical approach: method-based and process-based.

Method-based orientations utilize conventions and influences from the teacher's chosen genre or approach (such a Kodaly or Orff), which in turn inform the constraints or parameters used. There is creative choice available to the participants but practically, only within the available specified parameters. For example, methods that use Orff-designed instruments give example limit pitch choice, as they are pentatonic instruments. A reason for utilizing this type of instrument is that pentatonic instruments playing together sound consonant and so produce a pleasant-sounding product, as seen in a study with fifth-grade children (Beegle, 2010). The process in method-based approaches is often clearly delineated and sequential; every step has to be mastered before moving to the next as seen in a study by Kratus (1991): 1. Exploration; 2. Process-oriented improvising; 3. Product-oriented improvising; 4. Fluid improvising; 5. Structural improvising; 6. Stylistic improvising; 7. Personal improvising. One potential reason for creating a learning path such as this is to cope with curricula that demand substantive assessments; another reason could be to aid teachers who are unsure about their own improvising skills by providing a clearly defined path.

Process based orientations of teaching improvisation focus on developing the musical material the participants themselves create. The teacher can then negotiate parameters through scaffolding the contribution of the child while attempting to preserve their creativity. Good examples of this are in the work of deVries (2005), who took his son's improvising as a 'point of departure', (Nettl & Russell, 1998 p. 72) to guide development of vocal skills and though this, understanding of musical concepts. Similarly in Young's (2003) research with preschool children, participatory adults responded sympathetically to children's improvisations with the intent to structure and further develop exploratory play. It is important to appreciate that in these two examples the teacher is working one-to one with children, and this ratio is common amongst other studies of this type. An exception was found in Young (2008) where pairs and trios of children improvised together without a teacher and their collaborative mechanisms (non-verbal communication) were examined. Therefore, there is a scarcity of work examining improvising with larger groups of preschool children (n>4).

4.2.3 The Scottish Context for 3–18 Education

As I will consider my music workshops in an Activity Theory framework, I now turn to contextualising the broader educational environment. In Scotland there has been recent educational reform with the Curriculum for Excellence (CfE) being implemented in most Scottish schools from 2010. The curriculum aims 'to help children and young people gain the knowledge, skills and attributes needed for life in the 21st century, including skills for learning, life and work.' (Education Scotland, 2016) It was developed after a National Debate to address the lack of coherence in different educational stages (from ages 3–18), to prepare pupils for modern life, and, to give more them more choice (Education Scotland, 2004). Four capacities were proposed to frame children's personal development and prepare them to manage a changing and challenging world. These are described as confident individuals, successful learners, effective contributors and responsible citizens (Education Scotland, 2004).

CfE has been seen positively with the potential for progressive child centred methods of teaching (Priestley & Biesta, 2013). The philosophy underpinning CfE is seen as 'implicitly socio-constructivist' (Priestley & Biesta, 2013 p. 45) and offers potential teaching methods such as 'scaffolding' a term Wood, Bruner & Ross (1976) developed from Vygotsky's (1978) concept of Zone of Proximal Development. For the teacher, there is more agency in creating the curriculum content to meet the specific needs of their environment, (Biesta, Priestley & Robinson, 2015). Criticism of CfE has focused on the lack of knowledge content, and the perceived vagueness of prescribed experiences and outcomes (Biesta et al. 2015; Young, 2010). As well as this, there is a greater work burden on teachers than in the previous curriculum, as they have to tailor lessons to the specific educational requirements of each class. In addition, some writers have questioned the values that underpin the four capacities and the overarching purpose behind those values. For example, the purpose of these capacities may not solely be to encourage the development of the person but also be designed so that governments can benefit from developing these attributes in a future workforce (Watson, 2010).

In CfE, the Early Years curriculum for music emphasises participation, exploration and personal expression over learning measurable music skills, the two specific outcomes for Early Years learning in music follow:

> I have the freedom to use my voice, musical instruments and music technology to discover and enjoy playing with sound and rhythm. EXA 0-17a
>
> Inspired by a range of stimuli, and working on my own and/or with others, I can express and communicate my ideas, thoughts and feelings through musical activities. EXA 0-18a

I will now outline the theoretical and methodological tools I have used to analyse my empirical work.

4.3 Theoretical and Methodological Tools

4.3.1 Activity Theory as an Analytical Framework

As the aims of my PhD work were exploratory, I felt it necessary to explore the conditions of the workshops, to gain a rich picture of the qualitative features of the historical and cultural environment and the ways in which I operated within it. For this reason, I chose Activity Theory (AT) as it provides meta-perspective on complex situations (Engeström, 2014). Activity Theory (AT) originated from the work of three Russian researchers Vygotsky, Leont'ev and Luria and is a theoretical framework that can be used to analyse how activities within a practice, an activity system, are shaped by its cultural context, (Bakhurst, 2009). This can be understood in the elements of the hierarchical roles of participants, rules of the environment and larger communities (Engeström, 2014). The final, crucially important element, tools, can be take form in physical tools which are used in the activity such as a paintbrush, paints and canvas or symbolic tools such as concepts or images to guide the choices available to the participants of the activity.

AT has been used by researchers in pedagogy to analyse how separate elements are realised in educational activities, for example, a study looking at the tools by which scientific concepts were taught to preschool children, also revealed personal beliefs and cultural influences affected the teachers' use of these tools (Sundberg, Areljung, Due, Ekström, Ottander, Tellgren, 2016). In music education pedagogy, AT was used in a study by Johansen (2013) to explore instrumental practising and dimensions of student values within the activity and the larger cultural and historical context both of Jazz, and a formalised curriculum.

AT has been used to analyse many work situations as well as pedagogy, including business and hospitals (see Engeström, 2008) therefore, to contextualise the research questions in this chapter I will now describe the various elements of Engeström's (1987) AT diagram (Figure 4.1) with reference to pedagogical orientations adapted from Hardman (2007) and Sundberg et al. (2016).

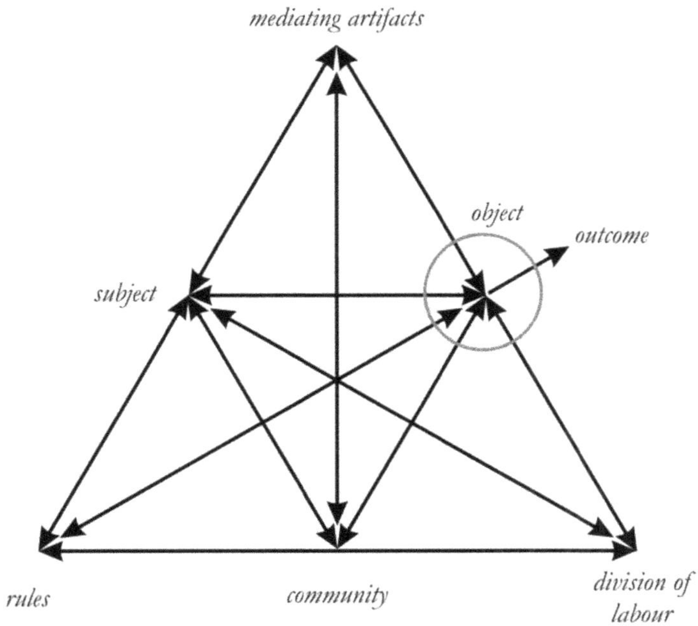

Figure 4.1 Diagram of Engeström's (1987) activity system.

Subject: The subject is the individual or group who acts in the system and whose perspective shapes the activity (Engeström, 1987). In Hardman's (2007) exploration of the AT elements, she refers to the teacher having "epistemic assumptions" about how knowledge is transmitted and gained and that these assumptions affect how the tools are used. For example, if a teacher believes that children learn though first experiencing and then discussion with a more experienced other, for example, Vygotsky's (1978) concept of Zone of Proximal Development, they will use tools in a different way than a teacher who believes children learn passively.

Object: Historically there is uncertainty about a clear definition of this element due to historical theoretical divergences; a description is these can be found in Kaptelinin, (2005). For clarity, this chapter will consider the object of the activity "problem space' as seen in Hardman (2007) which is transformed into outcomes through a range of tools which can be physical or symbolic (Engeström, 1987).

Tools/Mediating artefact/instruments: This element, can take the form of physical objects such as musical instruments or symbolic tools such as an image or verbal instruction. In educational settings, symbolic tools have taken the form of guiding questions designed to develop the children's understanding of their learning experience (Sundberg et al. 2016).

Rules: This element is concerned with the norms, conventions and guidelines for interaction in the activity system (Engeström, 1987). In Sundberg et al. (2016), rules were the guiding principles by which the teachers created the desired ethos for learning preschool science. Similarly, Hardman (2007) interprets rules as the "norms, conventions and social interactions which drive the subject's actions in the activity" (p. 77). In this way rules can be specific to the task and/or the teacher.

Community: A community comprises the wider circles which influence the object and has its own divisions of labour and responsibilities (Cole & Engeström, 1993). This element considers the immediate community of the participants of the activity (the teacher and the group of children) but, in a pedagogical setting, can also encompass wider communities such as the whole school and both local and national education systems (Hardman, 2007).

Division of labour: This can refer to a horizontal distribution of tasks or vertical divisions in hierarchies or power relationships (Engeström, 1987). In Hardman (2007), power relationships between teacher and pupils are described either as symmetrical or asymmetrical with children having agency to direct aspects of the activities in the symmetrical relationships, and very little or no agency in asymmetrical relationships.

4.3.2 Research Questions

The two research questions I will consider in this chapter are as follows, firstly, what was the educational outcome in my improvisation activities? Secondly, what tools were used in mediating these outcomes?

4.3.3 Novel Constructs

The two novel constructs of were created through my reflection on my understanding and experience of my own background as a free improvising musician and are as follows:

Creative Musical Agency (CMA): The child creates novel musical material independently and executes this in the group improvisation.

Socio-Musical Aptitude (S-MA): The child creates a musical response with reference to another child's musical idea in the group improvisation.

My aim in creating these constructs was to provide flexible and authentic constructs that could function as giving educational purpose to the activity of improvising c.f Biesta (2009). I also had the intended outcome of creating an experience of learning music in a group, emphasising the creative and social aspects of music making, rather than focusing on improving technical aptitude. The process of analysing key features of informal musical genres or styles with the purpose of bringing this to an formal educational setting has been explored by other educational researchers working in pop music (Green, 2002) and collaborative composition (Thorpe, 2015). Thorpe describes this process as 'back-engineering pedagogy' (2015, p. 164), however it is important to appreciate that in both contexts; students were collaborating towards a piece of music that had to be recognisably in a style within the popular music genre.

4.3.4 Methods

4.3.4.1 Study design

As stated previously, the work presented in this chapter is from the second of two cycles of Action Research carried out in two Scottish nurseries in 2015/16. A cycle comprised twice weekly, six-week programme of improvisation workshops with the aim of developing my two new constructs, CMA and S-MA and to refine the workshop strategies. In both nurseries, the children participating were in their preschool year of Nursery education, aged between 4 years 1 month and 5 years 3 months. Informed consent was required from parents for both their child's participation and for the workshops to be filmed. As well as this, I sought the children's verbal assent before every workshop. If the child did not wish to take part in a workshop, they were allowed to resume their everyday nursery activities without negative consequence.

4.3.4.2 Data gathering and analysis

1) Talk data from workshops

I videoed every workshop in Cycle 2, totalling 6 hours and 32 minutes of video data, and transcribed all of the verbal utterances from myself and the children which happened in the workshops. These were transcribed verbatim and analysed using Thematic Analysis following guidelines from Braun and Clarke (2013).

2) Music improvisation data

The improvisation sections from the workshops were sampled for further analysis, totalling 83 minutes for Cycle 2. To adequately investigate the children's improvisations, a multimodal approach to analysing the video data was chosen as the most appropriate to capture the detail and nuance of this context. A study by Korkiakangas, Weldon, Bezemer & Kneebone (2014) provided a relevant approach to adapt, as their work examined interactions between members of a surgical team (n=6), and, the researchers had to make analytical choices about which modes to transcribe, as not all modes of communication are equally important in a goal directed work or learning situation. Therefore their approach was considered the most useful in considering the context of examining key interactions in a mid-sized group of participants. Their coding strategy was adopted for my study and took the following steps:

1. observe data with an open mind
2. note down patterns in interaction which emerge
3. create categories that the patterns reflect
4. group categories and compare different incidences, from this the definitions of the categories will be developed
5. find the best examples of each category to examine in finer detail

I transcribed four modes of communication, verbal, music, gaze and gesture and then revisited the music mode for further refinement. Within the scope of this chapter, it is only possible to consider the music mode and a description of the further coding now follows.

I coded CMA events as one in which a child initiated a new musical idea which was qualitatively different from the existing music texture and noted the musical parameters on which this occurred. An example of an event I coded as CMA follows:

> Workshop 1, Cycle 2 – All of the children were playing a rhythmically entrained piece of music (for 30s) until Christine started playing substantially slower and louder than the others. Christine's action of playing both slower and louder was noted as a CMA event on two musical parameters: tempo and dynamics.

An event was coded as an S-MA event if a child was observed to change their playing or singing to match another child's on one or more musical parameters, for example:

> Workshop 6, Cycle 2 – Tess had a big drum for this particular section of the workshop. The teacher invited the children to play using the "just play" instruction and the improvisation began with another child, Jane stroking a cimbala (a small sting instrument, similar to a small dulcimer) very quietly. Tess then began playing her drum by scratching the surface very gently with a circular motion. Therefore Tess related her dynamic level (quiet) to Jane's dynamic level and played her instrument in a way which achieved this (i.e. by scratching it rather than hitting it). Therefore this musical event was coded as S-MA on the musical parameter of dynamic.

3) Teacher interview data
Two teachers from the nursery were interviewed using protocol for semi-structured interviews from Willig (2001). This method of data gathering has flexibility in that initial questions can be modified in light of participant responses. The topics for exploration in the interviews were to explore the teachers' beliefs and attitudes towards teaching, and ways in which they approached managing and facilitating group learning. The interviews were also analysed using Thematic Analysis.

4) My reflexive data
Auto-ethnographic data included the following: transcribed voice memos recorded as soon as I could manage after the workshops, with the purpose of

capturing my initial thoughts and feelings; written field notes from later in the same day of the workshop and, written reflections on informal conversations with teachers which happened throughout the 6 week programme.

The talk data provided information for the tools, subject and rules elements of the following activity systems. The video data provides information for the tools, object and outcomes elements of the following activity systems, and the teacher interviews informed the environment element. My own reflexive data provided information for the subject, tools, division of labour and rules elements. The next section will critically appraise the elements in two key activities in Cycle 2 of the workshops and examine the relationships between the different elements, thus gaining a multidimensional picture of the setting. Implications from the results will follow in the discussion section.

4.4 Results

4.4.1 Workshop Activity 1: Descriptive Improvisation

In this activity, I asked the children to suggest ideas on which to base their improvisation. Common suggestions included representing nature and playing 'happy' music. The following activity triangle (Figure 4.2) represents the descriptive improvisation, *star music*, which was a popular and enduring

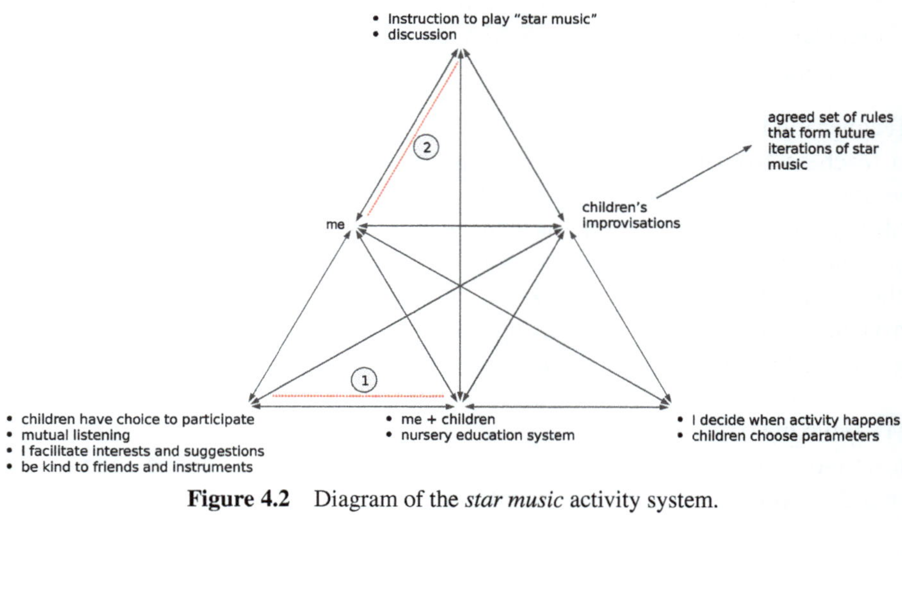

Figure 4.2 Diagram of the *star music* activity system.

activity in Cycle 2. As part of the workshop design I would engage the children in a discussion about the pieces afterwards, asking them open questions such as "What did you think about that piece?" with the aim of drawing the children's own views out and helping them build on their experience.

Subject: In this activity I am the subject, with the epistemic position to view myself as an experienced other and act as a creator of effective scaffolds for the children's learning.

Object: Children improvising with the purpose of creating *star music*.

Outcome: Agreed set of rules for creating future *star music* pieces. Table 4.1 shows the musical parameters that the children were able to manipulate to

Table 4.1 Table of agreed parameters for *star music*

Category / Parameter	CMA	S-MA	*Star Music*
Tempo	Child sings or plays a new tempo to the group	Child alters speed to match tempo of another child	slow tempo within range of 60–70 bpm
Dynamics	Child introduces a new dynamic which is louder or quieter than the rest of the group	Child alters their playing or singing to match dynamic of another child	Within range of pp (very quiet) to mp (medium quiet)
Articulation	Child instigates music which has a different articulation to the rest of the group (e.g singing short notes when the rest of the group is singing long notes)	Child matches their articulation to another child's. E.g singing short notes after another child has proposed this musical idea	Both long and short notes, but legato only (smooth notes)
Pitch	Child initiates a different pitch than the rest of the group	Child matches (or nearly matches) pitch of another child	Relatively high pitch for all children
Arrangement	Child plays starting and stopping	Child B starts and stops with child A who proposed arrangement idea	Children could start and stop as they wished – but there had to be a constant stream of sound

create *star music* according to their rules. Children could be creative and initiate new musical events (coded as CMA) or responsive (coded as S-MA) within the specific dimensions of the parameters for *star music* (final column).

Tools: Physical tools were the children's voices and instruments; symbolic tools were instruction to play *star music* and discussion after the piece. The discussion was both facilitated by myself and independently initiated by the children, most often correcting a perceived deviation from the rules for example "you played too loud, that's not star music" (quote from Tess, one of the children).

Rules: My rules for this activity were as follows: children have the choice to participate the workshop and in discrete activities and I will always aim to facilitate their interests and consider all suggestions from children. In Cycle 2, I had asked, during one of our discussions if the children had any suggestions for workshop rules and they decided on 1) be kind to friends and 2) be kind to the instruments.

Environment: I considered the environment to be on 3 levels, first the immediate setting of myself and the six children in the workshop; the next level is the other children and staff of the nursery and finally the wider educational environment of the local education authority functioning in a national setting.

Division of labour: At the beginning of the 6 weeks of workshops when beginning descriptive improvisation, I would take suggestions about what they wanted to describe through the improvisation. The rules of these improvisations coalesced through different paths; children decided the parameters of dynamics and pitch, in child-led conversations. I facilitated discussion about the arrangement parameter, with the aim of having the children think about whether they wanted to play all the time. Tempo and articulation were not discussed as these were parameters were stable and enduring through all iterations of this activity.

4.4.2 Tensions in Star Music

Tensions in the system are notated in the activity system above with a dotted line and a number. I will now examine each in turn.

Tension 1) There was a difference in expectation between the nursery teachers and myself about the children's participation. I had a rule that the children only participated if they wanted to, therefore, at times a child would sit out for a workshop activity. If the child did not want to participate at all they were free to go back to their usual nursery activities. This happened infrequently (4 times in 12 workshops). At times the nursery teachers expressed their opinion that they thought the children should all participate all the time. Varying reasons were given, for example "they shouldn't get a choice"; "it's so good for him, I don't want him to miss out" and "if the other children see he's getting to pick and choose, they'll all stop". These reasons may reflect the demands on managing groups of children to focus and learn. Also, there may not be adequate staff of facilities to accommodate children working by themselves.

Tension 2) was between myself as the subject and the constraints of the piece (tools). This arose when I wished to highlight and develop interesting musical choices made by the children, which didn't fit into the agreed rules. I felt internal conflict between different "selves", firstly as a teacher working in a socio-cultural ethos and as an improviser. My "improviser self" to new musical initiatives presented by the children but my "teacher self" balanced this out with not wishing to change the rules of the piece that the children had helped create.

4.4.3 Workshop Activity 2: Free Improvisation

In the workshop activity of free improvisation, the children were instructed to "just play" and we discussed the improvisation afterwards.

Subject: In this activity I felt it important to give as much freedom as possible to the children and not impose my own aesthetic on the children's improvisation.

Object: Children's group *free improvisation* with the purpose of developing CMA and S-MA.

Outcome: table of results for free improvisation – build repertoire of parameters, framed as CMA or S-MA. Table 4.2 shows what the group as a whole achieved over the 6 weeks. It is important to note that not all children used the last three parameters – alternative vocalising, sung material and spoken word. Four out of six children used all of the parameters and the other two children did not use the last three parameters.

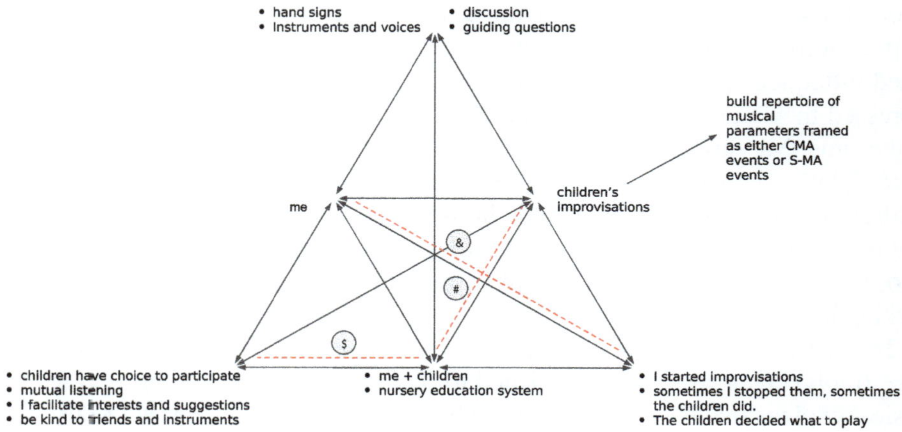

Figure 4.3 Diagram of activity system for *free improvisation.*

Table 4.2 Table of musical parameters used in *free improvisation*

Parameter \ Category	CMA	S-MA
Tempo	Child sings or plays a new tempo to the group	Child alters speed to match tempo of another child
Dynamics	Child introduces a new dynamic which is louder or quieter than the rest of the group	Child alters their playing or singing to match dynamic of another child
Articulation	Child instigates music which has a different articulation to the rest of the group (e.g singing short notes when the rest of the group is singing long notes)	Child matches their articulation to another child's. E.g singing short notes after another child has proposed this musical idea
Signs	Child initiates hand signs with the effect of changing the music – e.g long note sign when the group are singing short notes	Child responds to hand signs shown by another
Pitch	Child initiates a different pitch than the rest of the group	Child matches (or nearly matches) pitch of another child
Arrangement	Child A starts and stops playing or singing	Child B starts and stops with child A

Body percussion	Child makes a percussive action on, or with body (e.g claps hands, hits floor).	Child imitates percussive action of other child
Alternative vocalising	Child makes a sound other than speaking or singing with their voice (e.g altering timbre to be squeaky or growly)	Child imitates alternative vocal sound of other child
Sung material (in an instrumental piece)	Initiates sung words or small sung fragments of melody	Joins in with sung words or sings own words or melody
Spoken word	Initiates spoken single words or phrases	Copies spoken word proposal

Tools: Physical tools were the instruments and children's voices. Symbolic tools were guiding questions I asked the group about the improvisations, for example, questions about what they played and why they made musical choices, as well as what they found interesting in the improvisation. The children also used hand signs in the improvisations which at the effect of instructing other children to start or stop, or to play long or short notes. Without my input or suggestion, these hand signs were repurposed from warm-up exercises into the improvisations, by the children.

Rules: My rules for this activity were as follows: children have the choice to participate the workshop and in discrete activities and I will always aim to facilitate their interests and consider all suggestions from children. In Cycle 2, I had asked, during one of our discussions if the children had any suggestions for workshop rules and they decided on 1) be kind to friends and 2) be kind to the instruments.

Environment: I considered the environment to be on three levels, first the immediate setting of myself and the six children in the workshop; the next level is the other children and staff of the nursery and finally the wider educational environment of the local education authority functioning in a national setting.

Division of labour: I started free improvisations with an instruction to "just play", sometimes I would stop the improvisations. Children chose what they played and sometimes when to stop.

4.4.4 Tensions in Free Improvisation Activity

The tension marked as 1) was the same as descriptive improvisation. Tension 2) was identified from conversation with the children's teachers where they expressed uncertainty about what the children learned through free improvisation. This was accounted for as shown in an extract from my transcript from a conversation with Shona, one of the teachers from the cycle 2 nursery.

> The things you are doing with the kids... I'm not quite sure what it's getting at. But I'm not musical at all; in fact, this is not a musical nursery actually. They seem to be having fun though.

Shona had observed two workshops near the start of the cycle (workshops 3 and 4) where the children were experimenting with tempo, dynamics, articulation, signs and pitch in both vocal and instrumental improvisations. Although the children had both CMA and S-MA events, the improvisations did not have a strong melodic content in the way that a child's song has. A recognisable example is "twinkle, twinkle little star" which has a simple repetitive melody. The children's free improvisations did not repeat simple melodic patterns and this is a possible reason why the teacher did not 'get' the activity.

Shona then offers a possible explanation- "I'm not musical at all" which is an interesting point as nursery teachers use song every day in circle time to learn about numbers, animals and many other subjects. She reiterates this point more strongly by saying the whole nursery is not musical. My interpretation is that Shona's positioning of herself as unmusical is in relation to her knowledge of my background as a professional musician.

The final tension, 3), in the activity system of free improvisation was between the subject and division of labour. At times I stopped the children's improvisations for different reasons, sometimes I felt the music was getting too loud, that some of the children were hitting the instruments as hard as they could, with the effect of upsetting the other children in the group. At other times I stopped the group as they had played a static texture for a long time, my reason for stopping here was to begin a critical discussion with the children about the music they were creating, I thought they would find it easier to discuss music that had stable features, rather than try to talk about a changing situation. I felt ambiguous about stopping the children's improvisations, even thought I had good reason to, as I felt a conflict with my ideal position, as described earlier as allowing the children as much freedom as possible.

4.5 Discussion and Conclusions

In this section, I will return to my research questions and discuss the implications of my results with reference to relevant literature.

4.5.1 What Was the Educational Outcome in My Improvisation Activities?

As seen through Figures 4.2 and 4.3, each activity system had a different outcome, therefore the improvising in the object or "problem space" Hardman (2007) functioned differently in each activity, which aligns with MacDonald et al.'s (2012) assertion. The difference in the improvisations' functions can be further appreciated in considering where the divergence occurred in the two activity systems. The first difference to consider was in the position I described myself adopting in relation to each activity. In *star music* I described my participation in helping the children discuss and create the rules which formed the framework of future iterations of the piece. The children decided two musical parameters and I directly guided them to decide on another, therefore my decision about the level of my involvement was influenced by my principle to act as a scaffolder to the children's experiences. My position as a teacher in this activity involved me listening and allowing the children to negotiate with each other to create their own rules and if negotiations broke down or the children were quiet, I stepped in to a more active role. Contrastingly, in the *free improvisation* activity, my position as a teacher was to allow maximum freedom to the children which had the consequence of affecting the following discussion, which I will discuss further in the next section.

In Thorpe's (2015) study, she applied Reinharz's (2011, 1997) concept of different research 'selves' into a music education context. These selves were described as *research*, *brought* and *situational*. This is particularly pertinent for my work as the concept acknowledges the performer/teacher's professional background and identity in the *brought* aspect. As seen in the Sundberg et al. (2016) study, my personal beliefs about the nature of particular activities affected my use of tools (in this case discussion) and thus, my epistemic position was different in each activity.

When comparing the two tables of musical parameters, (Tables 4.1 and 4.2) the children experimented with more musical parameters in a more open way in *free improvisation* when compared to *star music*. However an interesting difference was in the participation levels of the children, all of the children contributed to all of the musical parameters in *star music*, but not all children explored the full range of musical parameters in *free improvisation*. There is no talk data available to illuminate why this occurred, perhaps it could be attributed to fluctuating levels of engagement with the activity or a child being shy about proposing a new musical parameter. For this aspect of evaluation of the workshop activities, it may be useful to consider Amabile & Gitomer's (1984) argument that children's musical creative thoughts may not be realised and so the music created doesn't fully represent a child's understanding or imagination.

4.5.2 What Tools Were Used in Mediating These Outcomes?

The *physical* tools used by the children were the same for each activity, they used both their voices and classroom instruments to improvise with. The main difference between the activities was in the use of *symbolic* tools through the talk sections of the different activities following the improvisations. In *star music* the discussion focused on agreeing structuring elements for future iterations of the piece and in *free improvisation*, the discussion consisted of "guiding questions" which had the purpose of encouraging the children to think critically about what they had played. For this age group, the questions were very simple and importantly there was no right or wrong answer as the key part of the discussion was to encourage critical reflection on their own and each others' playing. The children used another symbolic tool in free improvisation, by repurposing hand signs, originally a warm up exercise. The hand signs became tools for changing the improvisations by children showing each other a new musical direction.

4.5.3 Conclusions

Other researchers looking at group improvisation identified similar concepts shared key features with my constructs of CMA and S-MA, for example Burnard (2000) identified leaders who 'defined the direction in which the others should move" and followers who were 'musically led and influenced'. This conceptualisation of improvisation is similar to my constructs may help enrich the experience of the children as they have constructs which help them explore the complexities of these two roles. There are also similarities with CMA and S-MA and Young's (2008) work, which describes adult's sympathetic responses to preschool child-led initiatives. It is important to note however that this work is with and adult playing with one or two children, and, the adults interactions with the child being influenced by the roles and responsibilities in music therapy of therapist to client. This does suggest that the constructs of CMA and S-MA align with other areas of music work which involves improvisation, and thus gives a potential conceptual reframing for improvisation in plural contexts.

I would argue that the constructs serve as a useful overarching purpose for teachers, to help with conceptualising what can develop in improvisation workshops. As seen in Table 4.1 the children developed CMA and S-MA in a varied range of musical parameters, which offers an appreciable outcome. These tangible outcomes have potential to challenge teachers' and parents' beliefs about children's creativity in music by offering an alternative way to appreciate their contribution.

Reflecting on these activity systems reveals the complex roles and negotiations I encountered when teaching improvisation to preschool children. Using Activity Theory as a meta-analytical tool also illuminated tensions when teaching both structured and open-ended creative activities in a group. Finally, I offer two novel constructs as an effective educational purpose for future music improvisation activities, and, as a method of framing the improvisations in a manner that enables both musical development and to expand children's own critical understanding of what they play.

Acknowledgements

Many thanks to Dr. Guro Gravem Johansen for comments on an earlier version of this chapter.

References

Amabile, T. M., and Gitomer, J. (1984). Children's artistic creativity: effects of choice in task materials. *Pers. Soc. Psychol. Bull.* 10, 209–215. doi: 10.1177/0146167284102006

Ashley, R. (2009). "Musical improvisation," in *The Oxford Handbook of Music Psychology*, eds M. Thaut, I. Cross, and S. Hallam (Oxford: Oxford University Press).

Bailey, D. (1992). *Improvisation: Its Nature and Practice in Music*. London: British Library National Sound Archive.

Bakhurst, D. (2009). Reflections on activity theory. *Educ. Rev.* 61, 197–210. doi: 10.1080/00131910902846916

Barrett, M. S. (2006). Inventing songs, inventing worlds: the 'genesis' of creative thought and activity in young children's lives. *Int. J. Early Years Educ.* 14, 201–220. doi: 10.1080/09669760600879920

Biesta, G. G. (2009). Good education in an age of measurement: on the need to reconnect with the question of purpose in education. *Educ. Assess.* 21, 33–46. doi:10.1007/s11092-008-9064-9

Biesta, G., Priestley, M., and Robinson, S. (2015). The role of beliefs in teacher agency. *Teach. Teach.* 21, 624–640. doi: 10.1080/13540602.2015.1044325

Beegle, A. C. (2010). A classroom-based study of small-group planned improvisation with fifth-grade children. *J. Res. Music Educ.* 58, 219–239. doi:10.1177/0022429410379916

Braun, V., and Clarke, V. (2013). *Successful Qualitative Research: A Practical Guide for Beginners*. London: SAGE.

Braxton, A. (1985). *Tri-Axium Writings*. Lebanon, NH: Frog Peak Music.

Burnard, P. (2000). How children ascribe meaning to improvisation and composition: rethinking pedagogy in music education. *Music Educ. Res.* 2, 7–23. doi: 10.1080/1461380005000440

Cole, M., and Engestrom, Y. (1993). "A cultural historical approach to distributed cognition," in *Distributed cognitions*, ed. G. Salomon (Cambridge: Cambridge University Press).

de Vries, P. (2005). Lessons from home: scaffolding vocal improvisation and song acquisition with a 2-year-old. *Early Child. Educ. J.* 32, 307–312.

Education Scotland (2004). *A Curriculum for Excellence – The Curriculum Review Group*. Edinburgh: Scottish Government.

Education Scotland (2006). *The Purpose of the Curriculum*. Edinburgh: Scottish Government.

Education Scotland (2016). *What is Curriculum for Excellence?* Edinburgh: Scottish Government.

Engeström, Y. (1987). Learning by expanding. an activity-theoretical approach to developmental research. Helsinki: Orienta-Konsultit Oy.

Engeström, Y. (2008). *From Teams to Knots: Activity-Theoretical Studies of Collaboration and Learning at Work*. Cambridge: Cambridge University Press.

Engeström, Y. (2014). *Learning by Expanding: An Activity-Theoretical Approach to Developmental Research*. 2nd Edn. Cambridge: Cambridge University Press.

Green, L. (2002). *How Popular Musicians Learn: A Way Ahead for Music Education*. Aldershot: Ashgate.

Griffin, P. E., and Care, E. (2014). *Assessment and Teaching of 21st Century Skills: Methods and Approach.* Dordrecht: Springer.

Hardman, J. (2007). An activity theory approach to surfacing the pedagogical object in a primary school mathematics classroom. *Outlines: Critical Practice Studies*, 9, 53–69.

Hickey, M. (2009). Can improvisation be 'taught'?: A call for free improvisation in our schools. *Int. J. Music Educ.* 27, 285–299. doi:10.1177/0255761409345442

Hickey, M. (2015). Learning from the experts: a study of free-improvisation pedagogues in university settings. *J. Res. Music Educ.* 62, 425–445. doi:10.1177/0022429414556319

Heble, A., and Laver, M. (2016). *Improvisation and Music Education: Beyond the Classroom.* London: Routledge.

Johansen, G. G. (2013). *Learning from Musicians better than Me. The Practice of Copying from Recordings in Jazz Students' Instrumental Practise.* Oslo: Norges musikkhøgskole.

Johansen, G. G., Larsson, C., MacGlone, U., and Siljamäki, E. (2017). "Expanding the space for improvisation pedagogy: a transdisciplinary approach," in *Proceedings of the 22nd Annual Conference of the Nordic Network for Research in Music Education, Gothenberg 14–16 March 2017*, Gothenberg.

Kaptelinin, V. (2005). The object of activity: making sense of the sense-maker. *Mind Cult. Activity* 12, 4–18. doi: 10.1207/s15327884mca1201_2

Korkiakangas, T., Weldon, S.-M., Bezemer, J., and Kneebone, R. (2014). Nurse–surgeon object transfer: Video analysis of communication and situation awareness in the operating theatre. *Int. J. Nurs. Stud.* 51, 1195–1206. doi: 10.1016/j.ijnurstu.2014.01.007

Koutsoupidou, T. (2008). Effects of different teaching styles on the development of musical creativity: insights from interviews with music specialists. *Musicæ Sci.* 12, 311–335.

Kratus, J. (1991). Growing with improvisation. *Music Educ. J.* 78, 35.

Lewis, G. E. (2000). "Teaching improvised music: an ethnographic memoir," in *Arcana: Musicians on Music*, ed. J. Zorn (New York, NY: Granary books).

Lewis, G. E., and Piekut, B. (2016). *The Oxford Handbook of Critical Improvisation Studies.* Vol. 1: New York, NY: Oxford University Press.

Linson, A. (2014). *Investigating the Cognitive Foundations of Collaborative Musical Free Improvisation: Experimental Case Studies Using a Novel*

Application of the Subsumption Architecture. Doctoral thesis, Open University, Milton Keynes.

MacDonald, R. A. R., Wilson, G. B., and Miell, D. (2012). "Improvisation as a creative process within contemporary music," in *Musical Imaginations: Multidisciplinary Perspectives on Creativity, Performance and Perception*, eds D. J. Hargreaves, D. Miell, and R. A. R. MacDonald (Oxford: Oxford University Press), 242–255.

MacGlone, U., and MacDonald, R. A. R. (2017). "Learning to improvise, improvising to learn: a qualitative study of learning processes in improvising musicians," in *Distributed Creativity: Collaboration and Improvisation in Contemporary Music*, eds E. Clarke and M. Doffman (Oxford: Oxford University Press).

Nettl, B., and Russell, M. (1998). *In the Course of Performance: Studies in the World of Musical Improvisation.* Chicago, IL: University of Chicago Press.

Priestley, M., and Biesta, G. (2013). *Reinventing the Curriculum New Trends in Curriculum Policy and Practice.* London: Bloomsbury.

Reinharz, S. (1997). "Who am I? The need for a variety of selves in the field," in *Reflexivity and Voice*, ed. R. Hertz (Thousand Oaks, CA: Sage), 3–20.

Reinharz, S. (2011). *Observing the observer: Understanding Ourselves in Field Research.* Oxford: Oxford University Press.

Rose, S. (2012). *Articulating Perspectives on Free Improvisation for Education.* Unpublished doctoral thesis, Glasgow Caledonian University, Glasgow.

Sundberg, B., Areljung, S., Due, K., Ekström, K., Ottander, C., and Tellgren, B. (2016). Understanding preschool emergent science in a cultural historical context through activity theory. *Eur. Early Child. Educ. Res. J.* 24, 567–580. doi: 10.1080/1350293X.2014.978557

Thorpe, V. (2015). Assessing complexity: Group composing and New Zealand's National Certificates of Educational Achievement. (Unpublished doctoral thesis). Victoria University of Wellington, Wellington.

Vygotskiĭ, L. S., and Cole, M. (1978). *Mind in Society: The Development of Higher Psychological Processes.* Cambridge, MA: Harvard University Press.

Watson, C. (2010). Educational policy in Scotland: inclusion and the control society. *Discourse* 31, 93–104. doi: 10.1080/01596300903465443

Whitcomb, R. (2013). Teaching improvisation in elementary general music: facing fears and fostering creativity. *Music Educ. J.* 99, 43–50.

Willig, C. (2001). *Introducing Qualitative Research in Psychology: Adventures in Theory and Method*. Buckingham: Open University Press.

Wilson, G., and MacDonald, R. (2016). Musical choices during group free improvisation: a qualitative psychological investigation. *Psychol. Music* 44, 1029–1043.

Wood, D. J., Bruner, J. S., and Ross, G. (1976). The role of tutoring in problem solving. *J. Child Psychiatry Psychol.* 17, 89–100.

Young, S. (2003). Time-space structuring in spontaneous play on educational percussion instruments among three- and four-year-olds. *Br. J. Music Educ.* 20, 45–59.

Young, S. (2008). Collaboration between 3- and 4-year-olds in self-initiated play on instruments. *Int. J. Educ. Res.* 47, 3–10. doi: 10.1016/j.ijer.2007.11.005

Young, Y. (2010). Globalisation, knowledge and the curriculum. *Eur. J. Educ.* 45, 4–10.

5

Revisiting Japanese Multimodal Drama Performance as Child-Centred Performance Ethnography: Picture-Mediated Reflection on 'Kamishibai'

Hiroaki Ishiguro

Department of Education, College of Arts,
Rikkyo University, 3-34-1 Nishi-Ikebukuro, Toshima-ku,
Tokyo 171-8501, Japan

Abstract

Educators have long known that stories are an effective cognitive tool to extend the imagination (Egan, 2005). This paper discusses 'Kamishibai' (paper drama performance), a form of dramatic story-telling popular among young children in Japan. A Kamishibai story is composed of a series of picture cards, which teachers read aloud and use to involve children in the world of the story. Our team of researchers set up a play-based workshop called 'Playshop' (Ishiguro, 2017) in order to study the significance of play for children. Children who participated were expected to extend their zone of proximal development (Vygotsky, 1978) through rich bodily play experiences. We adopted 'the formative experimental method' (Vygotsky, 1960–1979) in which children are free agents in a roughly preset environment. This paper details two forms of Kamishibai we observed during Playshop: adult-initiated collaborative story-making and child-initiated story-telling. In the first, adult facilitators helped children assemble separate images (their own drawings) into a cohesive, shared story. The process of making a story shows children that the meaning of a story can change according to the author's way of composing it. The children experience this first-hand as they assume authorship in the making of their own story. The second, child-initiated Kamishibai, takes this a step further as individual children begin to internalise the collaborative process and make it their own. Both adult- and child-initiated Kamishibai proved effective ways to foster imagination and aid children's self-expression. The observations here thus have significant implications for the pedagogical application of Kamishibai.

5.1 Introduction

This study introduces a popular form of Japanese drama called 'Kamishibai', meaning 'paper drama' (de las Casas, 2006). It is 'a form of story-telling, or performance art that developed in Japan in the late 1920s' (McGowan, 2010). In this multimodal drama, actors use series of pictures to narrate or perform the story. Kamishibai is often used as a form of story-telling in kindergartens and nursery care institutions. The performer is usually an adult—a teacher, parent, or other facilitator.

Reading picture books aloud to children is known to have a positive effect on young children's academic performance (van den Heuvel-Panhuizen et al., 2016). The teacher usually involves the children in discussing the content of the book and supports their learning of vocabulary, conceptual development, comprehension, and content knowledge (Panayota and Helen, 2011). Kamishibai can be similarly effective for children's cognitive development. This study focuses on two forms of Kamishibai: adult-initiated collaborative story-making and child-initiated Kamishibai. The pedagogical application of Kamishibai for children to make their original story is not common practice in Japanese preschool institutions but a few practitioners do promote this activity in their daily program. Recently, Kamishibai is being recommended for school children even outside of Japan (de las Casas, 2006; McGowan, 2010).

5.1.1 Kamishibai as Performance Ethnography

Kamishibai is a kind of drama performance in which characters act out a story that is represented on a series of pictures. It is a popular form of performance art in Japanese early childhood institutions because it is inexpensive and easy to perform even for a non-professional. Kamishibai might be the first arts-based pedagogic activity that most Japanese children encounter. Adults perform ready-made Kamishibai for audiences of children. Arts-based pedagogy does not simply mean grafting artistic materials and technology onto pedagogy; it must also mean accommodating art's critical perspective. Art activities in school are not merely an effective sweet food to temper the more bitter subjects. Finley describes arts-based inquiry as 'performing revolutionary pedagogy' which demands transformation of the world of oppression, as follows:

> Arts-based inquiry, as it is practiced by academics doing human social research, fits historically within a postmodern framework that features a developing activist dynamic among both artists and social researchers. (Finley, 2005)

Art is considered as a critical public pedagogy, which considers 'how art can be used or created to help learners as creators of art, or as viewers to engage in a challenging dominant ideologies or to engage in social issues', is proposed by Zorrilla and Tisdell (2016). In Playshop, young children enact simple Kamishibai stories while adults facilitate (Ishiguro, 2017). The resulting Kamishibai is not an excellent artwork, but the process disrupts the traditional asymmetrical relation between adults as producers and children as consumers of art. Children are conceived as valuable producers who reflect on their shared experiences and collaborate in the making of Kamishibai for their enjoyment. In Playshop, we observed two forms of Kamishibai that emphasise the children's role as *participants* in pedagogical Kamishibai. The first was adult-initiated Kamishibai. An adult facilitator showed the children pictures they had drawn in previous sessions and asked for their ideas. It was a collaborative effort, and the medium of pictures drew out the children's reflection on their drama performance. The facilitator then assembled the children's ideas into one cohesive story. We employed collaborative Kamishibai as a way to enable children to comment on their daily activities. We found it was a way to advocate children's internal voices. Following the adult-initiated Kamishibai, during the conversational phase at the end, a few children began telling their own stories. They became active agents sharing their own ideas with others in what we've called child-initiated Kamishibai.

Kamishibai can be conceived as a kind of performance ethnography. Performance ethnography is the hybrid of performance studies and ethnography. McCall contends that a performance ethnographer is required to write a script, cast it, and stage it on the basis of his or her field notes (McCall, 2000). The resulting performance ethnography is called 'a performance installation' (Jones, 2002). In our study, the adult facilitators wrote collaborative field notes in order to share what was happening and how the participants felt and thought about their daily activities at Playshop. We also videotaped the activities and collected the pictures drawn by the children. These records proved important resources for understanding the children's perspective. We performed our own Kamishibai stories that resulted from the collaborative efforts of the children to reflect on the original Kamishibai story they had heard. 'By utilizing an experiential method such as performance ethnography, those who seek understanding of other cultures and lived experiences are offered a body-centred method of knowing' (Alexander, 2005). Kamishibai is thus a way to connect to children. It is very difficult to know what young children think and feel during their activities. Kamishibai-making and performance can mediate the children's experiences and make them more accessible to adults. The use of fiction in Kamishibai also differs from typical social science field notes in its particular ability to draw out children's emotional experiences. The children's voices collected during Kamishibai cannot be gained by interviews or questionnaires.

5.1.2 Brief Sketch of Kamishibai

Kamishibai is a popular cultural resource for telling stories to Japanese children in early childcare institutions. Its root was called 'Utushie' in the Edo era. (Minwaza, 2016). 'Kami' is the Japanese word for paper and 'shibai' refers to drama performance. Kamishibai, then, is a multimodal performance that combines picture cards with the performer's narration. Japanese pre-schoolers are very familiar with Kamishibai.

The Kamishibai I will be discussing here is called 'Kyouiku (pedagogic) Kamishibai' in contrast to 'Gaito (street) Kamishibai.' Pedagogic Kamishibai is usually performed in early childhood institutions, children's centres, and public libraries. Street Kamishibai is enacted by professional performers who sell candies to the audiences of children who gather to watch the performance (Tokyo City, 1935). Street Kamishibai requires a 'Butai (stage)' which is a wooden frame for the cards, but in pedagogic Kamishibai, teachers usually hold the cards in their hands instead.

Yone Imai, who established 'Kamishibai-Kanko-kai' (Kamishibai publisher) in 1933, is said to be the first person to use Kamishibai for pedagogical purposes. She was a Christian and used Kamishibai to teach Bible stories such as the Christmas story. Her activity is widely viewed as the inception of pedagogic Kamishibai (Bingushi and Taneichi, 2005). As television grew in popularity, street Kamishibai began to decline from 1953 forward (Shimokawa, 2002). As for pedagogic Kamishibai, however, it was referred to as one of the most important activities in the Ministry of Education's

Pedagogic Guidelines for Kindergarten. When the guidelines were revised in 1967, Kamishibai was excluded from the official school program. But it has remained a popular pedagogic tool for many practitioners in early childhood institutions.

Kamishibai uses sequential picture cards that make up a story. Each card has a picture on the front side and the story text on the back with a miniature replica of the front-facing pictures. However, the information on the back does not correspond to the front picture as the cards are stacked, but usually to the information for the next picture cards. This allows performers to keep track of which picture they are performing. The performers are recommended to perform the cards rather than reading them directly, like theatre. One of the most important actions in the telling of a Kamishibai story is the displacement of the paper cards (Horio, and Inaniwa, 1972). Through displacement, audiences can watch the physical transformation of the character or environment and, therefore, of the psychological transformation of the atmosphere. The action is very effective at getting the audience to empathise with the story. For example, when a performer reveals only part of the next card, the audience begins to imagine what happens, what the rest of the card will reveal.

The displacement procedure also involves the colour and composition of the picture. Figure 5.1 shows a scene from *Alice in Wonderland*. Alice is experiencing a transformation of size after eating a mushroom[1] (Paatela-Nieminen, 2008, p. 98). The background colour of half showing the bigger Alice is yellow and the background colour of the smaller Alice is red. When the performer begins to draw the card but stops before the entire

Figure 5.1 A card depicting the transformation of Alice in *Alice in Wonderland* (Paatela-Nieminen, 2008).

[1]I refer to *Alice in Wonderland* because of its international popularity. There are many Kamishibai that portray traditional Japanese stories.

picture is revealed, the audience can only see the bigger Alice. When the performer suddenly shows the left part of the card, the audience is surprised at Alice's transformation. Consider another example: each card represents one stationary moment of an event. If a performer shakes the card, the shaking may represent confusion.

Kamishibai enables children to understand a story from the perspective of an external narrator. Horio & Inaniwa argue that Kamishibai has a fundamental organizing structure, that is, a specific dramaturgy: 'hajimari' (arising), 'agari' (development), 'chouten' (peak), 'sagari' (calming), and 'owari' (conclusion) (Horio and Inaniwa, 1972). The text accompanying the pictures is composed of descriptive and conversational sections. The audience needs to be able to differentiate the characters' voices from the narrator's voice in order to understand the story in Kamishibai. This is often difficult for younger children. Furthermore, in conversational sections, it is difficult for them to differentiate which character is speaking. In other words, the audience is required to actively imagine the scenes and the story in Kamishibai in a different way than in a puppet show or a bodily drama play. These constraints may strengthen cognitive function for children. Other multimodal media are expected to provide the same effect in similar way. Animated cartoons and manga, which look similar to Kamishibai, are even more explicit in their identification of who is acting and speaking. For example, there are speech balloons in manga, and the actors' bodies move according to their dialogue (de las Casas, 2006). Japanese children generally are very familiar with Kamishibai, animated cartoons, and manga from their early years. Their experience of these media might lay a foundation for multimodal learning.

5.2 Materials and Methods

This study examines two kinds of Kamishibai—adult-initiated, collaborative story-making process, and child-initiated story-telling—collected from 'Playshop' (Ishiguro, 2017), a play-based workshop and experimental research site designed to study preschoolers' after-school programs and the psychological significance of play for child development. Practically speaking, it is a play place for participants ages 3 to 6, and uses a situation-based, negotiable curriculum. Playshop was held once a week for about two hours as after-preschool program in a kindergarten. The program began in the fall of 2003 in Sapporo, Japan. Graduate and undergraduate students participated as facilitators for their own research. Kindergarten teachers also participated as part of their in-service training, and observed aspects of their children they had not seen during the course of regular school activities.

Playshop's daily program included four phases: Picture-book reading or Kamishibai performed by an adult facilitator, the children's dramatization in collaboration with the adults, drawing pictures based on the story, and discussion about the day's activities. Kamishibai was intentionally used twice: First, when the teacher performed the picture drama, which was related to the theme of the day. The Kamishibai story reminded children of the previous day's activities and foreshadowed the next phase of the day: the dramatization. The Kamishibai in this phase was prepared by adults but often included the pictures drawn by children in the previous week. Then, the final phase—in which the children discuss the preceding phases—included Kamishibai-making activities. The facilitator encouraged the children to collaborate on a story using the pictures that they drew. In this phase, a few of the older children unknowingly began to perform their own Kamishibai, thus introducing a third, unanticipated phase of Kamishibai into the day's activities. The textual data shown in the following analysis was extracted from Playshop in the fall term of the 2004 school year. There were 25 children and 5 adults, including myself. I supervised all Playshop activities, especially design, implementation, and recapitulation. Three graduate students acted as assistant facilitators. The graduate student Sako (not her real name) sat in front of the class as a main facilitator while the others assisted her. One undergraduate student supported children as well. We obtained and documented research permission from all parents and the school principal with an investigation-request letter.

5.3 Results and Discussion

During the second week of the second term, 2004, I began to observe two kinds of Kamishibai play: adult-initiated and child-initiated. The drama we adopted for our theme that term was 'The Ant and the Grasshopper' from Aesop's fables (Ishiguro, 2005). The children were asked to draw their experience of drama in the third phase. We thought that they would be able to reflect on their experience of the drama through their drawings. When we found that some of pictures did not have any relation to the dramatization, we soon learned that their pictures did not relate to the day's activities. The children gathered in a circle to share their reflections during the last phase. We asked the children to give a verbal account of their experiences, but most of the younger children were not able to express themselves in this way. Even among the older children, only a few were able to comment on their experiences. We decided to ask the children about their pictures in order to help them express their experience. We could observe two kinds of improvised Kamishibai during circle time.

5.3.1 Collaborative Story-Making with Children's Drawn Pictures

Sako, one of our assistant facilitators, collected the children's drawings. Figure 5.2 shows the pictures she chose to help them create their story during the discussion phase. She showed the pictures to the children one at a time and asked them to explain each one. Some of the adults had drawn pictures with the children during the third phase, and these drawings were used as necessary to fill out the story. The children might not have been aware of the connection between their comments on the pictures, but in collaboration with the students, Sako tied all the pictures and children's descriptions together into one coherent story. The whole discourse using the four pictures indicated by Figure 5.2 is shown in Discursive Protocol 5.1. Table 5.1 indicates the main story lines which Sake drew from their discussion about the pictures. Each picture introduced a new piece of information

Figure 5.2 Pictures drawn by participants to reflect the story of the day.
Note: Picture 1 was drawn by an adult assistant facilitator.

Discursive Protocol 5.1 An adult-initiated story making process

1. Sako (showing the first picture [picture 1] to participants): All of you see this. How do you feel about it? What does the ant do? It looks like it is walking on the warm ground. **An ant was walking on the ground**.
2. Sako (turning to picture 2): Oh, the ant met a green insect. What is this? What insect is it, all of you?
3. Fen & Machi (children): Grasshopper!
4. Nasshi (child): An ant. A strange ant!
5. Sako: Is it a strange ant? Ok, **the ant met a green strange ant**. (Then, turning to picture 3): Oh, what is this? The ant met a green strange ant, and then what is it doing?
6. Dachi (child): He read letters that were written in the paper and he said them one by one.
7. Asa (child): Ra, Go, Ta, Ru, Tu, Ya, Kuwagata (a stag beetle). (All participants laughed.)
8. Sako: A stag? **A stag beetle came, too**.
9. Dachi: A stag beetle is not good!
10. Sako: Is it a problem?
11. Dachi: Yes. It was drawn with Satoshi's advice.
12. Sako: Ok, **an ant and a green ant were talking to each other.**
13. Dachi: No! It is not an ant but a grasshopper.
14. Sako: I see. Then, **an ant and a grasshopper were talking to each other.** What are they talking about?
15. Machi: [Whispers something to Maka (facilitator).]
16. Maka (on Machi's behalf): They said, 'Shall we play?' The grasshopper said, 'Shall we play?'
17. Sako: **The grasshopper said, 'Shall we play?'**
18. Maka (addressing Machi): How about the ant?
19. Dachi: It said, 'No!'
20. Sako: Ok. **The grasshopper said to the ant, 'Let's play', but the ant said, 'No'.** (Turning to picture 4): **And the grasshopper said, 'I'm sorry' and walked away.** Then it arrived here. What was it doing?
21. Dachi: Who drew this picture?
22. Fen: The ant is carrying.
23. Sako: The ant is carrying something. What is it carrying?
24. Dachi: Takoyaki
25. Sako: **The ant is carrying Takoyaki**.
26. Yashi: Does it eat the Takoyaki?
27. Sako (after showing the next picture): Then, oh, a grasshopper appeared again. What is the grasshopper doing?
28. Machi: It said, 'Shall we play?'

Notes: The names are anonymous. The sentences in bold became the main story line.

Table 5.1 The shared story, initiated by an adult facilitator

1.	**An ant was walking on the ground (picture 1 of figure 2).** 'Hajimari' (arising); introduction of protagonist and the initial state
2.	**The ant met a strange green ant (picture 2 of figure 2).** 'Agari' (development): happening
3.	**A stag beetle came, too (picture 3 of figure 2).** 'Agari' (development): happening
4.	**The ant and the grasshopper were talking to each other.** 'Agari' (development): happening
5.	**The grasshopper said, 'Shall we play?' (picture 4 of figure 2).** 'Chouten' (peak): peak of happening
6.	**The grasshopper said to the ant, 'Let's play,' but the ant said, 'No.'** 'Chouten' (peak): peak of happening
7.	**And the grasshopper said, 'I'm sorry and walked away.'** 'Sagari' (calming): end of the happening
8.	**The ant is carrying 'Takoyaki'.** 'Sagari' (calming): end of the happening

Note: 'Takoyaki' is an octopus pancake popular among Japanese children.

for the story text. Picture 1 introduced 'an ant'. Picture 2 introduced 'a green strange ant'. Picture 3 introduced 'a stage beetle'. A 'grasshopper' appears in line 14, without the prompting of a new picture, because when Sako said, 'an ant and a green ant were talking to each other' (line 12), Dachi (anonymous name) immediately said, 'No! it isn't an ant but a grasshopper' (line 13). That is, the grasshopper was already introduced as a green ant (picture 3, line 5).

The story they created, represented in Table 5.1, follows the formula of Kamishibai prescribed by Horio and Inaniwa, (1972) except for the conclusion: 'hajimari' (arising), 'agari' (development), 'chouten' (peak), 'sagari' (calming) and 'owari' (conclusion). Of course, this does not mean that the children were conscious of the formula. They might have simply been responding to each picture and then Sako connected them into coherent story. Some of the coherence may have derived from the fact that the pictures were drawn based on the cohesive story the children had already heard during the first phase. Sako scaffolded (Wood et al., 1976) in such a way that the children could be in touch with the dramaturgy even when they lacked the competence or consciousness to formulate it themselves. This textual formula is one of basic elements of literacy learning. Children can perceive the formula when they watch Kamishibai performed by an adult facilitator, but this may be an insufficient stimulus. The advantage of adult-initiated Kamishibai is that the children are directly involved in the process of making a coherent story.

This kind of active engagement may also be a key to inspiring child-initiated Kamishibai, as we will discuss momentarily.

Visual information can provide common ground and facilitate understanding. Vygotsky's discussion of pseudo concepts notes that a word meaning can be negotiated (in terms of its materiality) between a child and an adult (Vygotsky, 1934/1987). Thus a visual resource can promote discursive competence by giving children an image to attach their thoughts to. Sako used pictures to stimulate children's imagination and promote discussion. When she asked the children straightforward questions about the drama experience, she found it was difficult for them to talk about their play. But when Sako asked the children not about their experiences but about the *pictures*, the children were able to talk. The pictures serve as a point of reference. For example, Nasshi said, 'a strange ant' to describe the figure drawn in picture 2 (line 4, discursive protocol 5.1). Then Sachi said, 'Ok, the ant met a strange green ant' at the line 5. Dachi immediately disputed this, saying, 'No! It is not an ant but a grasshopper'. Here picture 4 functioned as a point of reference to help the children remember their shared experience of drama play. Multimodal performances, such as Kamishibai, give children a chance to enjoy the story even when the younger ones cannot exactly understand it only through text.

Sako picked up the children's ideas by asking, for instance, 'Is it a strange ant?' after Nasshi's utterance as 'A strange ant!' (line 3) and composed them into one coherent story. The children talked more about their own ideas in this discourse than they did during the discussion phase at the end of the routine. Sako often repeated the main story line so that children could follow the story that they were making. In this way, the adult facilitator advocated for the children's voices. Another facilitator, Maka, sat among the children so she could pick up their small voices. Some children are not comfortable talking to groups but will tell their ideas to others on a one-on-one basis. Machi whispered to Maka (line 15) and Maka spoke on Machi's behalf: 'They said, shall we play?' The grasshopper said, 'shall we play?' (line 16). In this way, Sachi and Maka provided the scaffolding on which the children built their Kamishibai story.

5.3.2 Child-Initiated Kamishibai

Children often imitate adults' behaviour. Vygotsky points out that a child's imitative behaviours indicate their upper line of proximal development.

Children may not be able to do certain things alone, but they can do them with the aid of adults. Vygotsky explains this in the context of play as follows:

> Looking at the matter from the opposite perspective, could one suppose that a child's behaviour is always guided by meaning, that a preschooler's behaviour is so arid that he never behaves spontaneously simply because he thinks he should behave otherwise? This strict subordination to rules is quite impossible in life, but in play it does become possible: thus, play creates a zone of proximal development of the child. In play a child always behaves beyond his average age, above his daily behaviour; in play it is as though he were a head taller than himself. As in the focus of a magnifying glass, play contains all developmental tendencies in a condensed from and is itself a major source of development. (p. 102) (Vygotsky, 1978)

When the children who participated in Playshop shared their experiences through collaborative Kamishibai-making, some of the children went beyond the collective activity. A few of the six-year-olds unknowingly began to perform their own original Kamishibai by combing their own drawings with the pictures of their peers. This behaviour can be conceived as the internalization of the collaborative activity. Discursive Protocol 5.2 shows the case of Satoshi, who—among the three children who performed their own Kamishibai—seemed to enjoy themselves the most.

Some of Satoshi's language represents conventional story-telling phrases, like '... in a place, there was a grasshopper and a stag beetle' (line one), which echoes the classically familiar 'once upon time in a land far away'.

Discursive Protocol 5.2 Satoshi's Kamishibai

1. Satoshi (after the cheering and clapping of the participants): In a place ... (Then, louder): In a place ... there were a grasshopper and a stag beetle. ... The grasshopper already—the grasshopper ... [unintelligible].
2. Unidentified child participant: Too fast!
3. Satoshi: I see.
4. Satoshi: The grasshopper became a larva from an egg ... and ... the ants and the stag beetle together ... the monster's ... the monster also ... together ... [unintelligible] bore eggs and ... They felt easy ... all of them ... very ... when the spring comes ... when the spring comes ... all of them tried to find foods.

Note: An ellipsis "..." indicates a pause on the part of the performer.

He also organised his story according to the compositional formula of Kamishibai: 'In a place, there were a grasshopper and a stag beetle. The grasshopper became a larva from an egg and the other insects bore eggs. When the spring came, all of them tried to find food' (portions of the story omitted to illustrate the story line). The logical connection between the parts was incomplete, but we can see that he made an effort to coherently organise his propositions. It is also evident that Satoshi was conscious of his audience, because he replied, 'I see' to the audience member's request 'Too fast' in line 3. This indicates that he did not consider textual composition alone, but was also conscious of the audience's perception of his performance. Textual cohesion and comprehensibility are very important in Kamishibai, and by engaging these elements, children are building a foundation for academic literacy in their future studies.

The children in Playshop were active producers of Kamishibai as well as consumers. They enjoyed learning the literacy skills that are woven into the formula of Kamishibai, including the children could not verbalize the compositional formula or speak conventional expressive words well. Most expressions in storytelling are attributed to written words. Written language thus has a character akin to that of scientific concepts, which derives the concrete from the abstract. An abstract word is acquired at first without a concrete referent but subsequently comes into the sense that it implies in any context (Vygotsky, 1934/1987). Vygotsky emphasises that scientific concepts are learned in collaboration with adults, especially teachers. Wertsch comments that 'scientific' concepts for Vygotsky are apt to be understood as 'academic' or 'scholarly', and that 'Vygotsky saw the relation between conceptual discourse and the social institution of formal instruction' (Wertsch, 1990). Vygotsky points out the weakness and the strength of the scientific concept as follows:

> The weakness of the scientific concept lies in its verbalism, in its insufficient saturation with the concrete. This is the basic danger in the development of the scientific concept. The strength of the scientific concept lies in the child's capacity to use it in a voluntary manner, in its 'readiness for action'. This picture begins to change by the 4th grade. The verbalism of the scientific concept begins to disappear as it becomes increasingly more concrete. (Vygotsky, 1934/1987)

The formula and conventional words are set in place first by imitative performance and then children may begin to incorporate them into their everyday conceptions, fostered by the imitative experience. This is how Vygotsky puts it: 'the scientific concept blazes the trail for the everyday concept. It is a form of preparatory instruction which leads to its development' (Vygotsky, 1934/1987). Even if children play in their bodily drama, they are not always aware of what they are doing and they do not reflect on it by themselves. Adult collaboration (asking questions and connecting the dots) helps the children understand their experience. On the flip side, if children only watch Kamishibai and perform it with bodily dramatization, they might fall into verbalism because the experience depends so heavily on spoken and written language. In Playshop, children watched Kamishibai, dramatized it, illustrated their experiences, and made and performed their own Kamishibai reflecting the original story. The compound activities of the Playshop program promote the intersection between concrete experience and linguistic experience. Vygotsky notes that the intersection between the two concepts usually occurs at around fourth grade. But the present study might indicate that the intersection can and does happen much earlier. This might depend, however, on the quality of play collaborated on with an adult. The time period should be verified in further studies. Play and learning activities with Kamishibai can nonetheless make a good learning environment for child development.

5.4 Conclusion

The experience of dramatization is a rich source for imagination. But it does not necessarily help children to reflect on their experience of the story. This paper details what Playshop taught us about Kamishibai and its application in pedagogy. Adult-initiated Kamishibai proved an effective way for the children to reflect on and express their experiences. Since we used the original story line and the drawings the children made in previous sessions, the story continued to develop week by week. This experience even led a few of the older children to mimic the role of the adult facilitator and perform their own Kamishibai. They became agents in what we have dubbed 'child-initiated Kamishibai'.

Kamishibai gives kids an important linguistic toolkit for understanding and reflecting on their experience. Those children who performed their own Kamishibai stories might also go so far as to consider appropriate performance expressions for their audience. They might not be able to critically reflect on their performance and story composition, but they may gradually become conscious of its effect on the audience by discerning discrepancies between their own Kamishibai-making experiences and the adult-facilitated activities. Knowing that their own stories do not have to correspond to any rigid facts also affords children the opportunity to create new meaning and experience genuine authorship in a way that is less intimidating, because it is facilitated by the visual framework. Moreover, kids enjoy the activity. Multimodal media such as Kamishibai offer the possibility to promote emotional ability in addition cognition for children. This study focused on only a few cases, so more cases will be considered. The applicability of the circulated program of bodily dramatization and collaborative Kamishibai-making will be further examined as well.

Acknowledgements

I would like to thank Sachiko Uchida and other colleagues who contributed to Playshop. I also express my gratitude to Dr. Robert Lecusay and to the reviewers for their thoughtful comments.

References

Alexander, B. K. (2005). "Performance ethnography: the reenacting and inciting of culture," in *The Sage Handbook of Qualitative Research* (3rd Edn), eds N. K. Denzin and Y. S. Lincoln (Thousand Oaks, CA: Sage Publications), 411–441.

Bingushi, K., and Taneichi, J. (2005). '*Kamishibai As Media in Early Childhood Care and Education—A Consideration Based on the History of Kamishibai.*' (St. Mary's College: Annual Report of Studies), Vol. 27, 53–67.

de las Casas, D. (2006). *Kamishibai Story Theater: the Art of Picture Telling*. Santa Barbara, CA: Libraries Unlimited Inc.

Egan, K. (2005). *An Imaginative Approach to Teaching*. San Francisco, CA: Jossey-Bass.

Finley, S. (2005). "Arts-based inquiry: performing revolutionary pedagogy," in *The Sage Handbook of Qualitative Research* (3rd Edn), eds N. K. Denzin and Y. S. Lincoln (Thousand Oaks, CA: Sage Publications), 681–694.

Horio, S., and Inaniwa, K. (1972). *Kamishibai: sozo-to-kyoikusei (Kamishibai: creation and pedagogic potentiality)*. Tokyo: Doshisha Publisher.

Ishiguro, H. (2005). "Development of imagination through drama play with adults," in *Proceedings of the Symposium 'Play, Drama and Life' at the First ISCAR Congress*, Sevilla.

Ishiguro, H. (2017). "Collaborative play with dramatization: an afterschool programme of 'Playshop' in a Japanese early childhood setting," in *Routledge International Handbook of Play in Early Childhood*, eds T. Bruce, P. Hakkarainen, and M. Bredikyte (London: Taylor & Francis/ Routledge), 274–288.

Jones, J. L. (2002). Performance ethnography: the role of embodiment in cultural authenticity. *Theatre Top.* 12, 1–15.

McCall, M. M. (2000). "Performance ethnography: a brief history and some advice," in *The Sage Handbook of Qualitative Research*, eds N. K. Denzin and Y. S. Lincoln (Thousand Oaks, CA: Sage Publications), 421–433.

McGowan, T. M. (2010). *The Kamishibai Classroom: Engaging Multiple Literacies through the Art of 'Paper Theater.'* Santa Barbara, CA: Libraries Unlimited Inc.

Minwaza (2016). Available at: http://www.minwaza.com/edoutsushie/edoutsushie-nitsuite/ [accessed, July 20, 2016]

Paatela-Nieminen, M. (2008). The intertextual method for art education applied in Japanese paper theatre—a study on discovering intercultural differences. *Int. J. Art Des. Educ.* 27, 91–104.

Panaycta, M., and Helen, P. (2011). Theory into practice, 50, 269–276.

Shimokawa, K. (ed.). (2002). *Kindaikodomoshinenpyō (Modern History of Children): 1868–1926.* Vol. 1 Meiji and Taisho era. Tokyo: Kawade Shobo Shinsha.

The department of social affairs of Tokyo City, *Kamishibainikansuruchousa* (A survey of Kamishibai in Tokyo). Tokyo City, 1935.

van den Heuvel-Panhuizen, M., Iliada, E., and Alexander, R. (2016). Effects of reading picture books on kindergartners' mathematics performance. *Educ. Psychol.* 36, 323–346.

Vygotsky, L. S. (1934/1987). "Thinking and speech," in *The Collected Works of L. S. Vygotsky, Vol. 1: Problems of General Psychology*, eds R. Riever and A. Carton (New York, NY: Plenum Press), 39–285.

Vygotsky, L. S. (1960–1979). "The instrumental method in psychology," in *The Concept of Activity in Soviet Psychology*, trans. and ed. J. V. Wertsch (Armonk, NY: M. E. Sharpe, Inc.), 134–143.

Vygotsky, L. S. (1978). *Mind in Society*. Cambridge, MA: Harvard University Press.
Wertsch, J. V. (1990). "The voice of rationality in a sociocultural approach to mind," in *Vygotsky and Education*, ed. L. C. Moll (Cambridge, MA: Cambridge University Press), 111–126.
Wood, D., Bruner, J. S., and Ross, G. (1976). The role of tutoring in problem solving. *J. Child Psychol. Psychiatry* 17, 89–100.
Zorrilla, A., and Tisdell, E. J. (2016). Art as critical public pedagogy: a qualitative study of Luis Camnitzer and his conceptual art. *Adult Educ. Q*. 66, 273–291.

6

The Accordion Book Project: Reflections on Learning and Teaching

Todd Elkin[1] and Arzu Mistry[2]

[1]365 Euclid Avenue, #103, Oakland, California 94610, USA
[2]576, 1st Stage Indiranagar, Bangalore 560038, India

Abstract

This chapter outlines the overarching ideas of the Accordion Book Project, an ongoing effort by Arzu Mistry and Todd Elkin to develop art-centered, transdisciplinary methods for teacher and student **practice, inquiry,** and **reflection**. The processes outlined stem from a core belief in supporting the development of agency in young people and teachers so they can become proactive drivers of their own practice (learning and art). The methods co-developed here bridge education and contemporary art practice. This chapter shares pictures of practice and a supporting educational philosophy involving the use of an accordion book as a hybrid of sketchbook, journal, field notebook, map and work of visual/conceptual art. The practices outlined here focus on: Developing a *responsiveness to the world* through 'deep noticing', capturing how 'things talk to you,' developing an *ongoing practice* using structures and strategies to support cycles of inquiry and action and modeling the use of accordion books as *sites of captured reflection*, revisiting previously documented ideas, coding them and triggering new explorations.

Keywords: Artistic practice, Teacher and student agency, Creative inquiry, Accordion books, Teacher inquiry.

6.1 Background

The practice of using handmade accordion books as a constructivist site for reflection, inquiry and praxis is something that we (Elkin and Mistry) chanced upon one summer at a professional development institute for teachers.

We had come together from our separate continents to co-teach at the institute and were eager to jump into dialogue. On the day before the institute started we were feeling especially inspired and full of ideas and wanted to capture and document our excitement at being in the moment of learning: The anticipation, the discomfort of the unknown, neurons firing with multiple inquiries and being in a synergistic environment where we were excited about the content being shared with us. In a moment of "necessity being the mother of invention", we grabbed what we had in front of us, a brown paper bag from a grocery store and began visually mapping, in words and in images, the twists and turns of the dialogue we had been having while walking that morning. At some point, we collaged an actual map of the area onto the brown paper because we wanted to annotate it with the exact locations and contexts where the ideas we were having had emerged. Later, we folded the bag back and forth into an accordion structure and this became the first of many accordion books we have since made together and separately (Figure 6.1). For the next five days we both kept adding to this base

Figure 6.1 Mistry and Elkin's first collaborative accordion book.

conversation in a variety of non-linear ways. There was an energy and excitement in this visual form of capturing, sorting, ideating and seeing ideas for emerging inquiries and experiments.

While others at that summer's institute captured content by taking notes on ruled pads or on their electronic devices, we were drawing, making symbols, capturing words, passing the book back and forth to identify connections, contradictions and questions and processing our learning in a completely different way. As artists and arts educators we were already accustomed to making meaning in ways that were non-linear and layered so this process came naturally to us. This first accordion book was a captured and built-upon constructivist dialogue between the two of us and served as a model for many of our evolving ideas about the potential for different kinds of exchange in teaching, learning and contemporary art practice. In addition, that first collaborative accordion book became a tremendously powerful artifact for us, an embodiment of the excited processing and learning we were doing that summer. It contained the DNA of so much of the teaching and learning we would do in the years that followed, both in terms of process and content. In subsequent years it became important for us to continue using inexpensive brown kraft paper for our accordion books, for two primary reasons: First, we liked that kraft paper is accessible and inexpensive. Second, we liked that the brown paper did not feel precious, giving us the freedom to capture uncertainty, false starts and confusion and to be messy if we wanted. Formally, using the brown paper gave us permission not to chase "good", because "as long as you stick to good you'll never have real growth" (Bruce Mao). We encourage others to use kraft paper too for these same reasons.

After that summer, we began together and separately to share this process of meaning making using accordion books with other teachers and youth both in our individual teaching contexts and in workshops we taught together across the U.S. We began to guide others through the emergent process of capturing what speaks to them and using that as an anchor to develop and pursue their individual inquiries. We also shared different ways of coding information and other visual and organizational strategies we had found useful in our own accordion book practice.

Very quickly, we found that many of the teachers we shared the process with eagerly took to it and after five years of sharing these accordion book forms and methods with others we began to hear from people far and wide, some were former students, some had taken a workshop with us and others had copied one of our handouts second hand without ever meeting us. We heard from people across the globe, from Washington DC, to Spain, to

Singapore and many other locations. Many of them said variants of, “I've used this with my students and I love it.” and/or “ I've taken your ideas and made them my own.” This sharing of strategies with other teachers and youth and then learning from the ways they have taken the form and adapted it to suit their own individual purposes and inquiries has become a central part of our own teaching, learning and inquiry. We began to feel strongly that there was something here. Why did teachers and students take to this? What did this practice have to offer, particularly with relationship to active learning? These have become driving questions for us. This chapter explores these and other questions and begins to present some of our findings.

The processing of learning is messy. There is emotion, absorbing and processing information, confusion, connections being made, digressions and reflection all often happening simultaneously. We found that it was possible to capture this embodied and whole experience of learning in our accordion books. Others have found their own very personal reasons to develop an ongoing accordion book practice.

6.2 A Proposition

The work we are presenting in this chapter is based on our core belief in supporting the development of agency in young people and teachers so that they can become proactive drivers of their own practices of learning, art making and any other transdisciplinary inquiries they are pursuing.

Agency consists of feeling empowered to take charge of one's own learning, giving one's self permission to be alert, to notice and inquire, and inquire about the world around us and within ourselves and act and reflect upon our actions.

Agency here is linked to personal and relational motivation to pursue learning and sits separate from a compliance based model of schooling (Calvert 2016) Although most learners, both teachers and students, exist within a structure of schooling, can their individual agency empower them to advocate for their best potential for learning? Agency has typically been positioned in counterpoint to structure. Below we detail further our notions of agency in relation to David Perkin's proactive learner (2009), Olivia Gude's artist in a democratic society (2009) and Emirbayer's and Mische's temporal and relational model of agency (1998). The process of using accordion books has helped us, our teachers and students find and grow this seed of proactive agency and develop pathways to an active practice of learning, doing and reflecting.

6.3 Research Questions

Our research questions have emerged from many years of deep engagement in accordion book practice. Our collaborative development of accordion book forms, methods and purposes, our individual accordion book practices and our work sharing these processes with teachers, artists and young people have led to several overarching questions. These questions drive our inquiry:

In what ways does Accordion Book Practice build teacher and learner agency?

What kinds of learning and thinking does Accordion Book practice support and make possible?

How have teachers and students taken Accordion Book Practice and made it their own and what can we learn from this?

How can Accordion Book Practice scaffold the cycle of Inquiry, Reflection and Practice?

In what ways are active Accordion Book Practice and Contemporary Art Practice interrelated?

6.4 What Is Accordion Book Practice?

Our use of accordion books **for capturing, visualizing, and building upon reflective thinking** emerges from a long lineage of hand crafted bookmaking. These many book-like forms include illuminated texts, codices, illustrated field notebooks, artists' books, sketchbooks, journals, and fanzines. We hold these many antecedents close to us and are inspired by them.

The accordion books and processes we have developed have some specific qualities: They are a self-created, self-initiated aesthetic space. Very simply, they are handmade, zigzag-folded paper (often inexpensive brown kraft paper) containing drawings, writing, collage; art ephemera either made or collected, color codes, visual metaphors, flaps extending vertically off of the top and bottom and within the book and pockets containing additional written, printed, collected or drawn pages (Figure 6.2). We have found that this simple structure allows for a great deal of flexibility and lends itself to circular, non-linear, and divergent/emergent thinking. These eye catching handmade books are a site for capturing observations, insights and fleeting thoughts, revisiting and building upon ideas, making connections between seemingly unconnected elements and arriving at new and unexpected visual and conceptual places, only to begin again, picking up threads in often non-linear ways. Accordion book practice makes noticing, thinking, learning and curiosity visible.

Figure 6.2 Form and structure.

Accordion Book process comes from a philosophy of learner-centered practice, where the learner (be they teacher, parent or student) is directing their own inquiries and as a result, their own learning. Accordion book strategies shared by us or adapted by students and teachers encourage the maker to construct and document their own thinking as they follow trains of thought, get lost, go beyond the obvious or given, break through conventional/consensus notions of 'truth'-in essence to arrive somewhere different than where they started and in doing so develop deeper understandings in ongoing often non-linear or emergent ways.

We see our **accordion books as maps** of our learning and teaching.

The following sections outline five core ideas of the practice as we have developed it. They are the key overarching concepts that inform important aspects of accordion book practice.

6.5 Core Idea #1: Mapping the Terrain: Exploring, Guiding and Getting Lost

"We organize information on maps in order to see our knowledge in a new way. As a result, maps suggest explanations; and while explanations reassure us, they also inspire us to ask questions,

> *consider other possibilities. To ask for a map is to say, "Tell me a story."*
>
> Peter Turchi, *Maps of the Imagination: the Writer as Cartographer* (p. 11)

One of the central metaphors that has emerged as an important component of our accordion book practice is Mapping. Mapping is a many layered and flexible metaphor that serves a variety of purposes for reflective practitioners. The "story" being told by the maps in our accordion books is the story of our learning, our processes and our transdisciplinary journeys.

To Map Is:
to document,
to give shape to,
to explore,
to guide,
to get lost,
and crucially, to *interpret* and make meaning.

Terrains Are:
Geographic
Temporal
Imagined
Cultural
Philosophical
Emotional/Cognitive
Conceptual
Qualitative
Quantitative

The shifting boundaries, borders, relationships, places, things, and ideas being mapped in accordion books are personal. They are chosen, curated, arranged, and presented by the cartographer/learner/artist. Maps in accordion books are literal, metaphorical, iterative, recursive, constructivist, and not necessarily linear, often circling back upon themselves and even contradicting ideas previously explored. Mapping is inherently about exploring and guiding (learning and teaching). It is the art of slowly unraveling/unveiling/making visible the things we are curious about.

6.5.1 Mapping

A Layered Approach: We have identified a layered approach to mapping in accordion book practice. Each layer is anchored by an intention. **Exploration mapping** consists of being *fully steeped* in the curiosity driven, emergent moment. An Exploration map is created *while* one is discovering a place, idea or experience. The learner documents the immediacy of the first person experience, capturing and collecting, exploring and getting lost. At the start of the journey there are no categories. **Reflection mapping** happens upon one's return from an experience. It taps into memories by sorting collections, organizing, and beginning to analyze the journey. This layer can paint an overarching picture by illuminating connections, systems, and relationships. **Inquiry mapping** evolves from the connections that surface in a reflection map. A place or idea grabs ones and pulls ones back to dig deeper. There is still an element of exploring the unknown, a set of unanswered questions, but now there is an inquiry that is driving one's map so that new patterns, deeper meanings, and new avenues for exploration and practice may emerge.

The Exploration mapping layer of accordion book practice is the portal into the other two layers and in many ways the foundation of the entire process. This approach to capturing visceral, immediate, emergent moments is in a sense the opposite of an analytical process. It is a strategy useful when one needs to quickly document a thought or feeling before the fleeting moment, epiphany or experience gets lost. We too often think that powerful moments are indelible, that they will etch themselves into our memory, and while this is true sometimes we cannot count on it being true all the time. This is why it is useful to capture and document important experiences as close to when they occur as possible. The practice of Exploration Mapping is an example of what art educator Sister Corita Kent lists in *Some Rules for Students and Teachers,* "Don't try to create and analyze at the same time. They're different processes." (Kent, 1968) The documenting (capturing and creating) of the important moment comes first, the analysis, connecting, categorizing and building new thoughts and new inquiries comes later. The immediacy of the capturing is driven as much by emotions as by cognition and is a good example of the ways author Catherine Elgin describes emotions and cognition being inextricably woven together in our lived experience of the world. In *Considered Judgment* (1996), Elgin argues that "emotions function *cognitively*, guiding and structuring our **patterns of attention**. Emotions "orient [us], focus attention, and supply grounds for classifying objects as like or unlike." (p. 168) Emotions are crucially important players, guiding us as we encounter both our internal and external worlds. The intense moments we document quickly in exploration mapping are concrete and visible manifestations of how emotion and cognition are bound together in our processes of making meaning.

6.5.2 Getting Lost and Seeing Anew

The mapping encouraged in accordion book practice is not just a representation of what is, but an intention to map that allows the learner to move from *exploration to inquiry* and in the process *get lost and see familiar terrain anew.* To truly map our experience of learning is to get lost, to be comfortable with uncertainty, and to hold and entertain dissimilar ideas simultaneously. Solnit in *A Field Guide to Getting Lost* (2006) articulates "that thing the nature of which is totally unknown to you is usually what you need to find, and finding it is a matter of getting lost." In order to get lost we must acknowledge a change in mindset as "getting lost was not a matter of geography so much identity, a passionate desire, even an urgent need to become no one and anyone..." (p. 5) Authentic exploration of anything, in contrast to the too often prescribed processes in classrooms inevitably involves uncertainty, messing about experimenting, wandering, being "in the weeds". David Perkins (2010), in *Future Wise* writes about the need to educate for the unknown, advocating that it "favors a vision of learning aggressive in its effort to foster curiosity, enlightenment, empowerment, and responsibility in a complex and dynamic world." (p. 23)

Curiosity-driven inquiry can lead us to complex, unruly, confusing and wholly unfamiliar spaces. We feel that these spaces are often the most important and powerful spaces to explore. Dealing with the complexity and uncertainty that confronts us when we are "lost" is a key component of self-driven practice. Resisting easy closure and being in a space of what David Perkins calls "optimal ambiguity" (Hetland, 2013, p. 97) can lead to unexpected discoveries, surprises and deeper dives into uncharted terrains. Artists are always in search of this edge between the known and the unknown and seem to engage with it with far less angst than others. There is something to learn from this desire for the unknown. Peter Turchi, in *Maps of the Imagination* writes:

> *"Artistic creation is a voyage into the unknown. In our own eyes, we are off the map. The excitement of potential discovery is accompanied by anxiety, despair, caution, perhaps, perhaps boldness, and,*

> *always, the risk of failure. Failure can take the form of becoming hopelessly lost, or pointlessly lost, or not finding what we came for (though that last is sometimes happily accompanied by the discovery of something we didn't anticipate, couldn't even imagine before we found it). We strike out for what we believe to be uncharted waters, only to find ourselves sailing in someone else's bathtub. Those are the days it seems there is nothing new to discover but the limitations of our own experience and understanding.*" (p. 13)

Designer/Educator Kenya Hara calls this process "Unknowing The World" (Hara, 2015) and calls the insights gained by allowing one's self to get lost in exploration "Ex-formation", conceived in counterpoint to Information. Ex-formation consists of insights that reveal to learners/artists "how little they know of the world" Because, says Hara "If you can figure out how much you don't know, the method by which you will know it will appear naturally." (p. 16) This kind of purposeful, generative getting lost has the potential to stand in stark and exciting contrast to the predictability of schooling, arts education and other disciplinary explorations.

Allowing oneself to truly get lost requires considerable courage and scaffolding. A step in this direction is to help young people and ourselves to be able to re-see what we have already been conditioned to see by our education or our upbringing. Maps and mapmaking can allow the **re-seeing** of previously "over-familiar" terrains in a multiplicity of ways. Maps illuminate connections and present both the big picture and the small details. They invite criticality, intuition, and the possibility of multiple readings of information and can uncover patterns and relationships and make systems visible. One goal of the different kinds of mapping in accordion book practice is to begin to see our "taken for granted" landscapes anew. Landscape researchers da Cunha and Mathur (2010) push against a certain conditioning that becomes embedded in our understanding of familiar places and acknowledge the near impossibility of imagining a place in any way that is different from the way it is embedded in our imagination. To see anew and to allow ourselves to get lost becomes a battle we have to play with ourselves. By visualizing, juxtaposing and applying coding strategies to both literal and metaphorical terrains we can shift the way we see and understand them in new and different ways. This process is akin to what Augusto Boal (2002) called "de-mechanization" a term the theater artist, educator and activist uses to describe the breaking down of our habitual, conditioned, reflexive ways of seeing and acting in the world. Boal accomplished de-mechanization through engagement in a variety of theater games. We think a similar process can occur when we apply different mapping strategies to our previously taken for granted terrains.

To be willing to be/get lost and have the confidence to let go of preconditioned ideas takes courage and a certain agency mindset, one that is steeped in a confidence of learning from the process and an excitement in learning for oneself. Perkins (2009) uses the term "proactive learner" to describe this taking charge as one explores unchartered territory.

> *"In general, proactive learners work to make the game worth playing for themselves, not depending so much on hit-or-miss inspiration from others nor on coercion with rewards and punishments. Teachers who encourage learners to take charge to some extent of their own motivation are helping them to develop autonomy as learners."* (p. 203)

Figure 6.3 Excerpts from Todd Elkin's accordion book.

6.6 Core Idea #2 Exchange as Art: What Happens in an Exchange?

> *"Talk to people you know. Talk to people you don't know. Talk to people you never talk to. Be intrigued by the differences you hear. Expect to be surprised. Treasure curiosity more than certainty."*
>
> Margaret Wheatley, Turning To One Another: Simple Conversations to Restore Hope to the Future (p. 145)

Learning is far from an individual pursuit and although in the above section on mapping we have focused primarily on the individual learner and their internal and external disposition to proactive learning, we also believe in learning as a social and collective exercise, a series of constant exchanges with others and the world around us. What exactly is taking place in an exchange? What changes in an exchange? And why are many contemporary artists organizing their work around different types of interpersonal exchanges? These questions seemed generative and important to us in this present historical moment in which we and many other educators are striving to create opportunities in classrooms and other learning spaces in which relational (Bourriaud, 1998) emotional, and interpersonal realms of teaching, learning, and collaborative art making can thrive. Claire Bishop (2011) in indicates that *"One of the main impetuses behind participatory art has been a restoration of the social bond through a collective elaboration of meaning."* (p. 12) For us and other arts and nonarts educators, the accordion book has become a flexible focal point for different types of powerful exchanges as it encourages acts of collective meaning making about the world we live in.

Why is the idea of exchange a generative lens through which to look at our practice as teachers, artists, and world citizens? The word exchange evokes diverse ideas ranging from things like love, gift, barter, argument, garage sale, to collaboration and globalization. We may not even be fully conscious of the many daily exchanges we make within the capital, social, interpersonal and redistributive economic systems that we live in. Exchanges of power in classrooms, in workspaces, and between individuals are happening all the time. Relationships in nature and between humans are sometimes mutually beneficial, sometimes exploitative, sometimes parasitic, and often transformative in both positive and negative ways. Looking at the various dynamics and qualities of these different types of interpersonal exchanges can reveal a lot about degrees of democracy, who is empowered, who is being disempowered, and what new forms and ideas are arising out of the exchanges. In their work on teacher agency, Biesta, Priestley and Robinson (2015) articulate an ecological view of agency that moves from seeing agency in isolation to looking at in relation to time (past, present and future) experience and possibilities. We share this here because this relational view of agency is linked to exchange and the motivations for learning that come from a variety of exchanges. In their definition of agency, Emirbayer and Mische (1998) describe agency as *"temporal-relational contexts of action" (p.* 970) again articulating the relational nature of agency to one's history, experience and interactions.

Classrooms, artistic collaborations, and other social settings can become experimental spaces for looking at and intentionally structuring different types of exchanges. Social Practice/Socially Engaged artists have been exploring this terrain for many years, creating relational situations or interventions in which exchanges between artists and audiences are foregrounded. Contemporary artists Rirkrit Tiravanija, Oliver Herring and Thomas Hirschhorn have created powerfully engaging works involving exchanges of food, conversation, artistic experimentation and moments of teaching/learning. Here, the role of the artist becomes that of a catalyst, setting off chains of interaction and inquiry. We find it exciting to think of educators and classrooms in this same way-as spaces of collective, collaborative, and relational possibility and action where we can look closely at **exchange as art practice**.

There are many possibilities for making accordion books a hub of exchange-centered interaction in classrooms, art spaces, between colleagues, friends and relative strangers. Structured protocols can be designed around different types of generative exchange. Here are some overarching categories around which exchange-based protocols and strategies can be designed:

6.6.1 Collaboration

In classrooms and other social learning spaces, accordion books have great potential for being at the center of meaningful collaborations. They can become flexible hubs for the exchange and influence of ideas on one another, moving each individual beyond what they are able to perceive/think/feel on their own. They can create opportunities for developing an open, appreciative approach to difference and otherness, helping to shift perspectives, change mindsets and come into contact with worldviews that are different than one's own. They can be catalysts for moving toward mutually built ideas, projects or products. Accordion Books can be a site where different types of collaborations can be planned, documented and enacted. These types of accordion book-centered collaborations shift learners away from the stereotype of the artist (or other disciplinary practitioner) as "lone genius" as well as moving learning communities away from education as individualistic pursuit/toward an acknowledgment of diversity and richness that comes from various kind of exchanges. Accordion book-centered collaborations can encourage perspective taking and acknowledging of differences. Students are thus able to become metacognitive about how exchanges influence them and their learning. In a conversation with Language Arts teacher Kristen Kullberg (July 2014. personal communication with Mistry.) we learned of her experimentation with accordion books with ESL students while the teacher and students read a text together. She created a shared accordion book with students and they used images and words to illustrate understanding of the text and created space for individual reflection and shared reflection. Kristen noticed a growing confidence in her students and tangible evidence of deeper understanding of the text.

6.6.2 Dialogue/Feedback

Dialogically driven classrooms and art spaces are potentially rich in both structured and free flows of communication. Accordion Books can be places where ongoing peer to peer feedback, critique and assessment can live. Learners can trade accordion books, and borrow or "steal" (with permission) ideas from each others' accordions. Ongoing dialogues between members of learning communities can be captured in accordion books and revisited/built upon over time. Because accordion books help make learning visible, there is an opportunity to see paths and traces of exchange when students view each others accordion books together and draw from each others' ideas.

6.7 Core Idea #3 Things Talk to Me: Being Alert to What Grabs Me

> *"I was paying so little attention to most of what was right before us that I had become a sleepwalker on the sidewalk. What I saw and attended to was exactly what I expected to see; what my dog showed me was that my attention invited along attention's companion: inattention to everything else."*
>
> Alexandra Horowitz, On Looking: Eleven Walks with Expert Eyes (p. 2)

The world is talking to us. How do we pay attention and talk back? At the very core of our accordion book practice is the act of **noticing**; being alert to how the world talks to us; developing a deep, nuanced sense of curiosity; engaging in "extreme noticing" then capturing and reflecting about the qualities and natures of the things we notice. For Students, artists, teachers and other disciplinary practitioners, developing dispositions associated with a deep attentiveness to the world is a prerequisite to the development of ongoing self-driven inquiries and a personal agentic practice.

As educators, artists and practitioners we have an explicit belief and a sense of trust that what catches and holds our attention is fundamentally valid and worthwhile of our focus. Unfortunately schools, particularly mainstream education both in India and the United States teach us that our observations are not as valid as the teacher's or the text. We as artists believe that one's internal attentional compasses are reliable drivers of interest driven learning, disciplinary practices and deeply personal reflections, and that we should nurture this compass. When we begin capturing, documenting, and being thoughtful about what we notice, we embark on a journey of many diverging paths. Following these paths, and creatively capturing, analyzing, and building upon the things we gather, is our practice. Olivia Gude (2009) describes this practice as being "intensely conscious of both inner experiences and of the prompting of the outer world—this heightened dual awareness is a defining characteristic of artistic process." Things are talking to us, and we are in dialogue with them. In this process, emotion and cognition are not dichotomous processes but are intimately joined. (Elgin, 1996). Thus, it becomes possible to develop a sense of personal agency by engaging in creative inquiry, meaning-making, and systemic thinking (Marshall 2014). It is worth mentioning that the proactive, self-driven, "noticing-centered" practice described above sits in contrast to business-as-usual compliance-centered versions of education and work life. Orienting ourselves to internal and external worlds represents empowering and democratizing notions of practice.

6.7.1 Grabbiness

In Todd Elkin's high school art classroom, in the courses he teaches for other teachers, and in his own transdisciplinary art and research practice the concept of **Grabbiness** has become a central and driving principal. Elkin defines the term this way: "Something is said to be grabby if it *catches*, and most crucially, *holds* your attention. Grabby things can be visual, conceptual, or both. Grabby things can be repulsive or attractive." Elkin uses the idea of Grabbiness and a closely related idea "Extreme Noticing" to center alertness, attentiveness, curiosity and the development of interest-driven, proactive inquiry in his pedagogy. Becoming alert and then looking for themes and patterns in the things we notice is the beginning of a journey. It is a habit you tweak and nurture into greater sensitivities until it informs your practice, your artwork and triggers new inquiries. We believe that supporting students in developing alertness to things in their worlds, both internal and external, sets them on a path toward their own personal practice. *Building alertness* to things and ideas is linked to the *inclination* to pursue an interest-based inquiry and the subsequent development of *skills* to execute that inquiry (Hetland et al., 2007) This is a cornerstone of our practice as educators and fundamental to accordion book processes.

Accordion books are spaces where the grabby things we notice when our external or internal worlds are talking to us are captured, documented, revisited and built upon. Those initial capturings and documentations, referred to above as "Exploration Mapping" are the first steps in developing an agentic personal practice. This is the crucial foundation upon which the other layers, Reflection Mapping and Inquiry Mapping can be built.

6.7.2 Revisiting Documentation

The artifacts/words/pictures we collect when we document a Grabby experience image or concept are not meant to be stashed away, never to be revisited. Carlina Rinaldi (2006) writes about the Reggio Emilia model of early childhood education, where documentations of experiences are brought back for rereading, revisiting, and the reconstruction of the experience so that they "intervene during the learning path and within the learning process in a way that would give meaning and direction to the process." (p. 25) Similarly, one of Todd Elkin's students spoke of the usefulness of accordion books in capturing/documenting Grabby things by referring to them to them as "external hard drives". Indeed accordion books can provide excellent 'back up' for our not always reliable 'human computers' as well as providing visually engaging foundations for ongoing constructivist learning and practice.

6.7.3 Unintentional, Intentional and Natural Grabbiness

One way to frame the idea of Grabbiness in a more systematic and perhaps more nuanced way is to organize the things that catch and hold one's attention into the categories of Unintentional, Intentional and Natural. Broadly,

Unintentionally Grabby things are just that, they are *not* purposefully and/or optimally placed in one's path in order to catch your attention. There is not a strategizing/animating intelligence that has placed these things in front of you with some kind of agenda behind it. Examples of Unintentionally Grabby things are the patterns of oil stains and cracks in a paved road, pieces of beach glass catching sunlight as they poke out of sand dunes, or the shapes that are formed in between the crossing wires from your lamp, modem and computer. Also in this category are juxtapositions of things not purposely placed together, like a child's shoe dropped in the gutter being partially covered windblown newspaper. The placements of Unintentionally Grabby things are more subject to the operations of chance than the other two categories listed above.

Times Square in NYC is in many ways the paradigmatic example of **Intentional Grabbiness.** Every square inch is packed with visual information, from gigantic animated billboards to scrolling LED news bulletins to Jumbotron video displays. Times Square is, like other major centers of 21st Century cities, a cacophonous symphony of purposefully placed visual/aural stimulation primarily designed to influence the spending habits of the viewer/consumer. Indeed, advertising, media, and other works of visual and popular culture shape our "noticing" muscles in strategic ways. In his seminal book, *Ways of Seeing* (1972) John Berger offers us a plethora of critical lenses to examine and understand the endless sea of imagery vying to intentionally grab and hold our attention, in some cases designed to influence us to spend or vote. Here Berger unpacks what he sees as one motivation behind the strategic and manipulative psychological techniques employed in advertising, which Berger refers to here as "publicity": "Publicity persuades us of such a transformation by showing us people who have apparently been transformed and are, as a result, enviable. The state of being envied is what constitutes glamour. And publicity is the process of manufacturing glamour." (p. 131) While advertising and other forms of market-driven visual culture are important examples of Intentional Grabbiness, there are others. Works of "Fine Art" also fall within the category of things that are Intentionally Grabby. While works of fine art occupy a complicated and somewhat slippery space within systems of commerce and culture, there is a sense within the fine arts (which can and must be looked at skeptically/critically) that the animating purposes and strategies visual artists use to intentionally catch and hold viewers' attention have much more to do with "purely" aesthetic (or anti-aesthetic which is an aesthetic of its own) and/or conceptual concerns and/or the attempt to give shape to internal and external 'worlds' than the "purely" market driven motives and techniques of advertisers. This Art/Commerce divide may very well be a complete fallacy (look at blue chip fine art auctions for evidence of this fallacy) but there has been a general and widespread contention, primarily born out of the history of artistic avant-gardes in modernist and postmodernist movements and reinforced in arts universities, that contemporary fine artists enjoy a greater sense of freedom to explore, experiment, offend and otherwise catch their potential viewers' attentions in unusual ways, ways untethered to and unconcerned with the concerns and strategies of the advertiser and the market. What complicates this already problematic space is the fact that advertisers/graphic designers (many of whom are art school graduates) continually appropriate the visual and conceptual strategies of contemporary visual artists so that a) it has become harder to tell the difference between "fine" and "commercial" artworks and consequently b) "fine" artists must work harder to differentiate themselves and their work from "crass commercialism". Another artistic subcategory of Intentional Grabbiness consists of visual works explicitly designed to directly persuade viewers in a political sense. These works, often labeled Propaganda have been ever present features in our visual landscapes almost since the invention of reproducible images. Makers of Persuasive Art/Propaganda use these visual works to convey strongly held opinions/beliefs about political issues and are trying to influence the viewer/audience to take some kind of action. The purpose of the artwork is to convince the viewer that this opinion is correct and/or that a particular action is called for. Persuasive Art often appeals to emotions, hopes and fears as well as moral beliefs. It is important for teachers and parents to support young people in developing a critical visual literacy in the face of the ever present surround of all types of visual culture coming literally from all directions.

The Final Category of Grabbiness, familiar to anyone who has been captivated by a sunset, a starry sky or a desert landscape is **Natural Grabbiness**. In his book *Truth Beauty and Goodness Reframed* (2011) Howard Gardner recounts that for many people, and even across cultures, notions of beauty are derived from qualities found in nature. (pp. 42–44) For whatever reason (evolutionary adaptation, neurological "hardwiring", cultural conditioning) things from the realms of nature seem strongly compelling to great numbers of people across the world. Indeed, the authors of this chapter each have our own separate corners of the natural world that "talk to us", that consistently catch and hold our attention. For Arzu Mistry it is plants, and after a lifetime of being drawn to them Mistry has become a scholar of plants. Mistry writes, "I've learned to read their clues and tricks for enticing me as well as all the insect, bird, and animal collaborators they lure and put to use." (Mistry & Elkin 2016, p. 11). For Todd Elkin it is Crows. After years of observing them and their behavior Elkin writes "I now care about them and am concerned for their well-being. I'm alert to and aware of both anecdotal and scientific evidence of their intelligence and abilities. I perk up when I see or hear anything about crows in print or online. My interest in them has become a strand of semi-formal research." (p. 42) For both of us, our accordion books have become the sites for our ongoing strands of research, musings, images and ever deepening engagement with trees, crows and many other things both natural and human made.

It is perhaps obvious but still important to note that the "things that talk to" different individuals and the reasons why they are Grabby are as various and diverse as we are as separate sentient beings. We are all informed by our own idiosyncratic sets of experiences, cultural lenses, points of view, frames of reference, biases, fascinations, pet peeves, traumas, memories and dreams. These factors shape not only *what* we are drawn to in our worlds but also *how* we interpret the things we find to be Grabby. A key element in determining both the what and the how of noticing and making meaning, as mentioned above, is our emotions, or perhaps more accurately the ongoing interplay between our emotions and our cognition. Far from taking an unreliable backseat to cognitive processes, our emotions play a crucial role in how we engage with and process our worlds. And so, to live is to be in a unique (to each of us) ongoing dialogue between the things that talk to us, our cognitively functioning emotions and the continuous meaning we make of it all.

6.7.4 Using Accordion Books to Capture the Dialogue

As we have described above, accordion book practice can be a layered process of mapping one's journey of exploring, reflecting and embarking on new inquiries. The world is talking to all of us twenty four hours a day (even in our dreams) and Accordion Books help us to document, organize, and analyze our ongoing dialogues with the world and most importantly make it possible to revisit, reflect and build upon them. David Perkins (2014) talks about the question of whether or not educational topics have an "afterlife" or "comeuppance", in other words, once topics are presented in classrooms, do they ever again "come up" in learners lives? Do they have a useful life *after school*? These two somewhat playfully humorous terms are another way for Perkins to ask the question "Is this topic relevant? Is it indeed worth learning?" (p. 52) Gardner (2011) Speaks of one of the three antecedents for determining something to be beautiful being "the impulse, the inclination, the desire to encounter again, to revisit" (p. 53) Here we can transpose the word Grabby with Beautiful. Grabby things are by their nature worthy of exploring further, our interest in them gives them an ongoing "afterlife" and accordion books are the space where that afterlife takes place and can be built upon. This is the practice of ongoing art-centered research.

6.8 Core Idea #4: What Would an Artist Do? Teachers and Students as Contemporary Artists

We think **contemporary artists are excellent role models for both teachers and learners**. They are proactive, self-driven free agents. They are "fired up", in that there is a palpable sense that they are compelled to do what they do, and have an extremely high level of engagement and sense of purpose in their work. As teachers, we would love nothing more than to help ignite this kind of drive and "purposiveness" in ourselves and our students. The sense of purposeful agency evinced in contemporary artists' practice is precisely what we have been referring to throughout this chapter as the set of dispositions we want to support as educators. In addition, the different types of idiosyncratic process-based practices of artists provide what we think are extremely useful models for teaching, learning and transdisciplinary inquiry.

An artist's work is driven by a variety of catalysts, but at its root, artists, like scientists, mathematicians, historians and other disciplinary practitioners are trying to understand the world. Toward that end they are proactive meaning makers in these ways: *artists synthesize, envision, reframe, catalyze, trigger, juxtapose, persuade, inform, translate, narrate, hybridize, analyze, bridge, connect, invent, explore, experiment, predict, play tricks, observe, critique, provoke, think systemically, inquire creatively, explore multiple perspectives and take part in conversations about the crucial issues of our time.*

Also:

Artists' work is research; an inquiry process that often results in new insight.
Artists are both explorers and guides.
Artists trust processes, have growth mindsets and embrace failures as unavoidable and instructive.
Artists understand that the world is an inherently transdisciplinary place, which means that they must adopt transdisciplinary approaches.

Artworks are catalysts for dialogue and the construction of meaning.
Artists give shape to internal and external worlds.
Artists are both critical thinkers and critical changemakers.
Artists perform functions that other types of practitioners do not. They affect transformations, operate simultaneously on intellectual and emotional levels, and dwell in liminal, interstitial spaces. They are shamans, tricksters, magicians, healers, and breakers of boundaries and taboos.

But where and how have the powerful dispositions of contemporary artists and the forms, methods and purposes of contemporary art practice stood in relationship to the goals/purposes and practices of progressive education?

There are many ways in which contemporary artistic practices include explicitly or implicitly epistemological, or pedagogical aspects. Occurrences of contemporary art, increasingly happening outside of museums and galleries, often feature different types of participatory, reflective and potentially transformative dialogic/interpersonal exchanges. In addition to exchanges of ideas and dialogue, of material goods and services or of collaborative actions, much contemporary art practice, either intentionally or not creates moments of teaching and learning. Artists such as Suzanne Lacy, Abigail DeVille, LaToya Ruby-Frazier, Oliver Herring, Rirkrit Tiravanija, Tania Bruguera, The Yes Men, Thomas Hirschhorn, the team of Allora and Calzadilla and many others have created interventions in social spaces or acted as catalysts in collaborative situations in which participants engage in various types of creative inquiries. These often interdisciplinary exchanges-sometimes resulting in the making of physical art pieces, sometimes consisting of ephemeral improvisatory or facilitated verbal dialogues share the quality of decentering the role of the artist, shifting the locus of creative inquiry on to other individuals or groups of participants. These shifts in dynamics are parallel to changes that have been taking place in progressive education circles, such as the Reggio Emilia focus on student centered inquiry or Eleanor Duckworth's Critical Exploration methodologies (2006) or indeed the central role of non-hierarchical dialogue in the pedagogy of Paolo Freire (1970).

It is tempting to trace the roots of the relatively recent reorienting shifts in contemporary art practice and their pedagogical bent- shifting away from the confined/confining spaces of galleries and museums and toward a variety of 'relational' social spheres-to the late 1950's when Allan Kaprow pioneered Happenings, artworks designed as socially constructed collaborative events. One could also go a bit further back to the early paradigm shattering and provocative work of Marcel Duchamp, whose Readymades, ordinary factory-made objects such as snow shovels, bicycle wheels, and most famously The Fountain, a urinal, upended notions of what could and should be called art and thus triggered all sorts of contentious dialogue and debate in and outside of the "art world" in Duchamp's era and beyond and for the first time ushered in notions of "art as idea".

Duchamp and Kaprow's provocations have reverberated through the decades and have inspired all sorts of conceptual practices and theories which in turn have triggered fresh rounds of debate around the increasingly blurry borderlines between 'art', 'life' and what defines an educational space. Indeed Allan Kaprow himself explicitly entered the educational arena in the late 1960's with Project Other Ways, in which Kaprow and educator Herbert Kohl took over a storefront in Berkeley California and engaged with public school students and teachers in a variety of workshops and semester long explorations. According to Kaprow, "Project Other Ways was intent on merging the arts with things not considered art, namely training in reading, writing, maths and so on." (Bishop, 1994, p. 84) In one notable Project Other Ways experiment, Kaprow worked with a group of Berkeley 6th graders who had previously been labeled as "hopelessly illiterate" and through a project in which the students photo-documented their communities and then annotated and wrote increasingly skilled descriptions and narratives about their photographs, Kaprow and Kohl facilitated a space in which this group of young people were able to, using artistic strategies, collaboratively give shape to ideas about their communities' heroes and villains as well as rewriting, illustrating and casting critical eyes on outmoded school readers containing problematic stereotypes and biases. Ultimately, through the art-centered interdisciplinary methods and works produced by the youth in Project Other Way, these 6th graders were able to show the teachers and administrators in Berkeley that they were in fact extremely literate after all.

The boundary pushing works of artists who make work today in the genres of Social Practice or Socially Engaged Art Practice are descendants of Duchamp and Kaprow as well as of the paradigm stretching, conceptual, street, feminist, and video/performance artists from the 1960's to today who have consistently sought to bring 'art' and 'life' together in meaningful ways to effect social change, spark insights and indeed to inspire learning.

6.8.1 Contemporary Art Practice/Pedagogic Practice/Accordion Book Practice: Challenges, Methods and Forms

Through intentional reorientations/shifts we've made in the way we frame and define what we do, the dividing lines between transdisciplinary teaching, learning, research and contemporary art practice are increasingly falling away. We no longer feel the necessity to compartmentalize what had previously been separate, albeit related undertakings. For us, there is increasing overlap between the heretofore separate aspects of our practice.

There have been criticisms to thinking about teaching and learning in this way. Over the past several years art historian Claire Bishop (2012) has written a series of works that critically push back against both Socially Engaged Art Practices and in particular "Pedagogic Art Projects". Although many of Bishop's critiques are valid, we feel there is considerable value in the educative qualities of contemporary artistic practice. For many years Elkin and Mistry have taught a course that shares the title of this section, *What Would an Artist Do?* In it we have endeavored to build a bridge between education practice and the work of participatory/contemporary artists and thus give teachers tools and insights about the connection between the two. The course gives teachers an interdisciplinary path into the strategies of artists and in particular, the way artists respond to the world around them. We share many examples of artworks dealing with important issues of our time. The artworks engage science, history, journalism, sociology, architecture and more. Using artifacts from contemporary issues, Mistry and Elkin designed a collaborative thinking process akin to the accordion book processes of exploration, reflection and inquiry mapping, coding and collaborative conversations in order to support teachers in seeing the interdisciplinary intersections in the content and practices of artists. Through orientations that focus on Deep Observation, Constructing Narrative, Persuasion, Perspective Taking, Experimentation and Intervention teachers are encouraged to develop rough drafts for contemporary art interventions. Both the processes in the course and the outcomes of a plan for a contemporary art intervention are designed to make visible to teachers the link between progressive education and contemporary art practices and the potential for creative agency in the classroom. The course models for teachers several possibilities for contemporary artistic approaches to addressing important specific and thematic issues of our time. We strongly believe that the forms, methods, knowledge, purposes, and contemporary artists' ways of being and doing are essential and core components of teaching, learning, and being in the world.

Accordion book practices are situated within an overarching set of process-driven methods, forms and purposes in which students and teachers are engaging in proactive interest-driven creative inquiry. Taken together these sets of methods and purposes form a pedagogy consistent with both the processes and products of contemporary art practice. Finally, there is a sense in which the processes of contemporary art inquiry are always epistemological, in that artists are engaged in a flexible and continually unfolding iterative dialogue in and through their work which is very much a self-directed process of learning. And while accordion book practices are very much process-centered, engaging in the gathering, sorting, mapping, coding, analyzing and building upon the inquiries in accordion book practice, there is an intention of action where learners/teachers/artists create, select, focus, frame and curate their content to present as works of contemporary art.

6.9 Core Idea #5: Getting Out of My Own Way: Trusting the Process and Resisting Closure

> *"Sometimes when I make work, there is a moment when what I want to make and what I make it with fuse in such a way that the piece begins, against my intention, to take on a form of its own. It is as though I am no longer the prime mover. At this point what is in front of me becomes as strange to me as I am essentially to myself. This is the point I am always trying to reach."*
>
> *Paul Chan, Selected Writings (pg. 189)*

There are many differing purposes and drives behind personal inquiries and reflective practice. While one motivation to inquire comes from curiosity about the world, other inquiries might come from an individual's desire to make one's practice better, to move beyond what we know (or think we know) or often, as we will discuss here, to get unstuck. Teachers as contemporary practitioners are constantly navigating the shifting grounds of their field, their own beliefs, experiences, insecurities, and personal practice. At any given moment in a teacher or learner's life there are forces that can easily lead to immobilization or becoming fixed in limiting ideas or ideologies. Teachers and students are continually subjected to ever-shifting cacophonies of educational paradigms, positions and mandates bombarding them with ideas, challenges, critiques, and memes. Sometimes these memes provide pithy, quick answers to the daunting and messy issues of our time. By distilling the complexity of our world, memes offer us for/against pro/con binaries and catchy quick fixes that run the risk of causing people to form instantaneous and exclusionary allegiances. Some memes can feel reactionary and reductionist, and are often false binaries which often foreclose the possibility of seeing multiple ways, perspectives and points of view. Learning that over-focuses on predetermined outcomes, "products" or notions of "excellence" can also cause us to feel stuck and uninspired, numbly marching towards foregone conclusions or rushing towards "definitive" closure. In this jostle, what happens to the "inner-selves" of teachers and students? (Narayanan, 2008) How can we persevere in our personal practices with all these outside forces, standards and mandates creating potentially paralyzing obstacles for us?

In addition, it is easy to feel overwhelmed by the pervasive "noise" in contemporary life which can interfere with our ability to make meaning to the point where suddenly we cannot "see the forest for the trees." We can become immobilized by uncertainty—so uncomfortable with not knowing that we ultimately quit trying. Or we feel stuck in certainty, which makes it seem like there is just one way to think or act which blinds us from seeing alternative ways. Whatever the causes may be, our classrooms, workspaces, and indeed our internal thoughts and feelings can too often feel like predictable, constraining, non-permeable "boxes" that we want to escape. The feeling of getting stuck, or being "in our own way", like failure, is an inevitable occurrence in all human practices. Finding ways to persevere through the different types of noise and moments of difficulty however is key to any thriving creative inquiry.

6.9.1 How Do We Get Out of Our Own Way?

We have found that this negotiation between personal process and the internal and external "noise" in our worlds makes having a personal reflective practice (accordion book practice) even more necessary. Through this ongoing iterative and often non linear process we can begin the journey of creating permeable and integrated learning; we can start to see continua rather than false binaries; to trust our process and thus become more comfortable with uncertainty; to be more alive and alert to the possibilities of a given moment rather than worrying about goals, deadlines and outcomes.

Many artists find it invaluable, when stuck, to tap into the dynamics of randomness or chance and thus release themselves from repetitive and/or stifling patterns. Artists Aris Moore says *"When I am stuck . . . I just search for excitement, but not too hard. It is when I find myself playing more than trying that I find my way out of a block."* (Krysa 2014, p. 22). Songwriter Bob Dylan throws I-Ching coins (2005) Brian Eno and Peter Schmidt created the Oblique Strategies: a deck of cards, pulled randomly, that offer ways out when an artist is feeling "blocked". Poet/singer Patti Smith consults tarot cards at decisive moments in her work. There are many other strategies shared by artists and designers to break creative blocks through the strategic use of chance, by tapping into hidden reserves of perseverance, by "letting go" or creating "generative limits". Here are a few examples:

Constraints: Abstract painter James Siena (2012–13) gives himself sets of directions for his artworks which he calls algorithms. Generative tensions arise between the rigidity of the rules he sets for himself and what his fallible human hand can actually do. Siena has found these types of constraints to be ultimately freeing.

De Mechanization: As mentioned above, theatre artist, activist, and political visionary Augusto Boal (2002) developed numerous games for actors and non-actors alike, and believed they could be used to transform and liberate everyone. Within his concept of **de-mechanization**, the mind is tricked by the action of the body to free itself from the patterns of our upbringing and the conditioning of social acceptability. "The process of 'thinking with our hands' can short-circuit the censorship of the brain." (p. xxiii)

The Paradox of Control: In his work on Flow Theory, Csikszentmihalyi (1990) shares eight components of the phenomenology of enjoyment. Of these, the "paradox of control" (p. 59) is particularly relevant here. He writes about the distinction between exercising control in difficult situations and being worried about losing control. When we become dependent on being able to control, we start to lose control. Order and control can become addictive and result in an unwillingness to engage with life's ambiguities. A conscious letting-go of the need to control can be a step toward feeling less stuck. Figure 6.4.

Stretch and Explore/Engage and Persist: In their seminal work resulting in the development of the Studio Thinking Framework and the Eight Studio Habits of Mind, Lois Hetland, Ellen Winner, et al. (2013) have proposed that the essence of an arts-centered education is not the art product, but rather the development of "artistic mind." Key to this development, and to the frame of this section, is the fact that artists are continually devising new ways to explore and reach beyond their habitual ways of doing and knowing. Rather than viewing mistakes and "failures" as dead ends, artists often see them as opportunities to grow. "Stretching and Exploring" are important parts of getting out of our own way, as is, crucially the inclination to "Engage and Persist". As art educator Corita Kent stated, "the only rule is work. If you work it will lead to something." (Rule 7). Hetland and Winner's identification of *Engaging and Persisting* as a crucial set of dispositions of the artist aligns with the ideas of Carol Dweck, who identified the phenomena of Growth

Figure 6.4 Excerpt from Jesse Standlea's accordion book.

and Fixed Mindsets. (Dweck, 2006) A growth mindset (as exemplified in an artist or learner Engaging and Persisting) sees learning, rather than being an "either you get it or you don't" proposition, as something in which there are multiple opportunities to deepen and grow understandings. In their book Studio thinking 2, Hetland and Winner also reference the work of Mihaly Csikszentmihalyi (1990) in describing the ways in which the high levels of engagement found in any self-driven artistic practice are "autotelic" (self-rewarding) a key feature in a state of "Flow" (p. 67) This type of intrinsic satisfaction creates in the engaged artist/learner an inclination to "push through" inevitable setbacks and failures. We have seen this deep sense of engagement in accordion book practice and believe that the ongoing and consistent capturing of all aspects of process (breakthroughs, challenges, failures) the "dailyness" of practice, can bolster growth mindsets conducive to engaging and persisting.

6.10 Pictures of Accordion Book Practice

As we have mentioned above, a great many teachers, artists and students have reported to us that they they have found using accordion books, in the ways we have been describing in this chapter to be an extremely useful, flexible and a multifaceted tool. For many, the accordion book has become a central hub in their various pursuits, woven/embedded into their teaching/learning/inquiries in ways they describe as expansive and generative. In the following section we will share the work of some fellow educators and students who have embraced the use of accordion books, initially through interactions with Elkin and Mistry (in both formal and informal settings/learning communities) and eventually as a regular/ongoing personal practice. Many of those who have continued to use accordion books beyond their initial interactions with us (as students or colleagues in the professional development sessions we have led for educators and artists) have developed their own very personal sets of strategies/forms and purposes for their work with accordion books, tailored to suit their own particular needs and goals as learners/teachers/artists. This proactive use of accordion books' various forms and methods is evidence to us that this practice has supported these individuals in being agents of their own learning/inquiry/artistic practice. Here are the stories of a few such individuals who have embraced accordion book practice as an important element in learning, teaching and/or artistic practice.

6.10.1 Derek Fenner: Things Talk to Me, Exchange, What Would an Artist Do?

Derek Fenner is a poet, educator, publisher, visual artist and researcher currently completing a PhD at Mills College in Oakland CA, USA. He has done extensive work with youth in the juvenile justice system and is currently leading ongoing professional development opportunities for teachers and youth leaders throughout the state of California and the U.S. as well as publishing collections of poetry through Bootstrap Press, the publishing house he co-founded in 2000. Derek had been familiar with the accordion book as a form in bookbinding and printmaking and was a student of different types of codices. He became reacquainted with the form through his work with the Alameda County Office of Education's Integrated Learning Specialist Program, where Todd Elkin had introduced accordion book practice as a common reflective/meaning making strategy amongst the teachers and learners in this sequence of three courses for educators based in the San Francisco Bay Area. Through his work in ILSP as faculty and a collaborative leader in the program Derek was able to develop related sets of accordion book strategies useful in the many different contexts he works, including his work as a poet and a doctoral candidate. Here Fenner describes some of the qualities of his own very personal experience using accordion books:

> "I used them myself way before bringing them into the classroom setting, and what I noticed is that it is a true representation. It feels

> more like a painting than a journal because it represents time in a way that painting would represent time. So it's not a snapshot. It's actually something that's a much slower process that's developed on the surface over a much longer period of time." (D. Fenner, personal interview with Elkin, July 20, 2017).

Fenner has noted a progression of personalization similar to his own happening amongst the teachers and youth, he has been sharing accordion book processes with. "What I love about them [accordion books] is that they're very much unique to the individual making it. And they may start off with just replicating ideas from other people, 'Oh, it's cool you put a pocket' or, 'Oh, it's cool you collaged,' or, "Oh, it's cool that you're coding it in this way."

> "When I started using it with teachers and students, all of a sudden it was after they got into it a little bit, their own aesthetics started to seep in and all of a sudden you could tell. Like, an accordion book halfway through, all of a sudden everything shifted. And they started off and it was all of the models that we were giving them or they were seeing from their fellow students, and then they ended up eventually ending, and most of them would then immediately start over a new accordion book because they had found their method."

Derek also found that the cultural lineage of accordion books, that they are descendents of Mesoamerican codices and other early book forms, to be an important fact to share with a group of student ambassadors he was working with in Southern California recently: "we talked about them being codices, where they're from, I kind of walked them through an uncovering of that. Like, 'What is this? What do you notice?' Using VTS [Visual Thinking Strategies] kind of showing them pictures from accordion books, that this is huge amounts of information, that it's the beginning of mathematics, all of these things and kind of showing them that this comes from the Yucatan and really helping them to connect with their own culture. Like, 'Oh, wow. These are the first books' in a lot of ways. They pre-date the printing press. So for them to then say, 'Now we're going make them,' they got so excited."

In addition to Derek's use of accordion books as a tool in his work teaching teachers and young people, he has found the practice to be useful in his own research work, specifically as an important source of ideas for the doctoral dissertation he is currently writing for Mills College. Derek asked his doctoral advisors if he could use his accordion books as primary sources

Figure 6.5 Excerpts from Derek Fenner's personal accordion books.

in his auto-ethnography and was granted permission. Fenner explained, "I wanted to include 6–12 pages from my actual accordion books-to interrupt the flow of the dissertation with the flow of my actual mind. And that's sort of the close read that I'm giving to it, is that, okay here's a great time to talk about race and trauma [major themes in Derek's work] but also here's a great time to talk about a process that's worked for me as a researcher. And also to represent my mind as closely as I can."

Lastly, Derek Fenner has developed a strategy in his accordion books centered around the use of quotations from poets, academics, artists and even colleagues. Fenner uses these quotes as catalysts, or generative "springboards for deep dives." Derek's accordion books are filled with quotations from widely disparate sources-the cumulative or synergistic effect is that "it starts to create a cacophony, or like a story. It starts to create the things that actually make me who I am as a creative person, as a writer, as a scholar, and I think that it gives honor to the lineage and that which came before. So it's been really important to me in terms of process." This strategy of channeling the words of others is central to both Derek's personal work as a writer and to his teaching. "So my notebooks have always been filled with the words of others because I, and again, I do this with the art-centered literacy stuff I do. It's really important for us to find ourselves in the words of others. That's a tool that builds empathy, which leads to solidarity." Out of this quotation-based meaning making process, Derek has developed another accordion book strategy for himself and his students that he calls "a reverse method of citation" Fenner explains, "We teach citation in one way in this country, or, in most places. And that's, "Come up with something that you want to say about this text, and then find the quote that backs it up. Well, what if we go into the text and we find something that moves us and then we say something about it? That's often more personal, more valid, and actually more interesting." Fenner's personalization of accordion book practice, his adaptive and innovative pushing the limits of the form in both his personal work as a poet/visual artist/researcher and in his work as an educator of youth and other teachers is an exemplar of the accordion book as possibility space in teaching, learning and artistic practice.

6.10.2 Caren Andrews: Mapping Terrains, Getting Out of My Own Way, Exchange

Caren Andrews is a visual artist and elementary school art teacher at an independent school in San Francisco CA. Like Derek Fenner, she was introduced to accordion book practice by Todd Elkin when they co-facilitated a course on assessment for teachers. Since then, Caren has developed a variety of innovative ways to use accordion books both in her work with students at the elementary school and with the teachers she continues to teach in the Integrated Learning Specialist Program in the San Francisco Bay Area. In her personal teaching practice, Andrews uses her own accordion book as a flexible tool to sequentially document the school year, house her ongoing overarching questions and document ongoing assessment "I also make my notes on students so it's where my basic assessment is. I record either behavioral or artistic breakthroughs or challenges in here." And so, for Andrews, the accordion book becomes a teaching tool, a way of mapping, organizing and making visible the terrain of a given school year for a given group of learners.

In addition, Caren uses accordion books with both adults and elementary school learners as a site for exchange

> "..the appeal to me for accordion books overall has always been that they stand up, that they break the two dimensional wall, so to speak, and they become a personal thinking wall. There are ways for one to be private about what's inside the books if one doesn't want to expose it to the general population. However, it's a great way for us to stand up our ideas and be able to look at them and exchange them. One of the strategies that's been great for me personally as an artist or as a teaching in the classroom studio setting is that we can steal like artists from one another. We can put the information out and without words look at each other's work very easily, and then find things that we like and integrate that into our own work as well as offer constructive criticism, feedback. "Have you considered?", "Would you try?", so that can happen from an idea point of view where I see an intellectual idea that I want to spiral out on, or a new way of integrating art materials into an accordion book." (C. Andrews, personal interview with Elkin, July 26, 2017).

For Andrews, the accordion book is a place for her young students to capture process. Here, Caren describes how accordion book practice is woven into her K-4th grade students' work " I start with accordion books in kindergarten, five year olds. I mostly work with accordion books with my kindergarten through fourth graders, the five to ten year olds. I use one accordion book per unit, and each grade level has I would say, on average, seven units that we cover per year. I build them for them in kindergarten when they first meet me, and then I teach them how to build their own afterwards using rhymes and storytelling to get them to get it internalized so that by the time they're in second grade, it's automatic. Usually it's earlier, but often they forget over summer and we have to go back through the folding in half and turning it into a W or wings or whatever it is that they've fixated on."

"The cover often is part of a two-minute warmup drawing that's connected to the curriculum. If it's about "how are artists scientists and artists at the

Figure 6.6 Excerpts from Caren Andrews accordion books.

same time?', then the cover starts to become a Venn diagram about what scientists are and what artists and where those crossovers are. Then inside, it becomes the research that we're doing about that project that's investigating that bigger question that we're getting smarter about. Again, at the end of the year, they aren't all connected together because of that communication piece, but they're all grouped together in the student's process folio that ends up going home at the end of the school year with instructions to have the child set up a gallery walk for the parents and have them walk through. I also have a script of questions for them to ask about the process so that again, it's not always just about which projects they liked the most and how it turned out beautiful or whatever it is. You're trying to train the parents into focusing on the importance of the thinking and developing a, using Kimberly D'Adamo's words, a 'thinking studio, not an art studio'." (Marshall, 2011)

Finally, Andrews tells of a particularly powerful exchange with one of her fourth grade students, whom we will call A. This anecdote, we think, illustrates how reflective engagement with accordion books can help learners *"get out of their own way"*:

> "A was one of those elementary students that if not engaged, would run around the room and poke other students, get in their business and start to irritate in order to get attention. I knew that about A. I'd been teaching her for at least four years. I gave her my accordion book for that school year. I asked her to look for two things. I gave her two sticky notes. (Figure 6.6) I asked her to list out *what did she see*, and I asked her to look for *what was missing or what was needed.* She sat there. I teach in 45-minute blocks, so this was probably a 20-minute block that she was engaged in my accordion book, really taking the challenge seriously. At the end, what she saw and the response to, "What do you see?" was *'life, art, a light bulb, telling a story, your passion, the brain, and how life is a big*

> *question.'* Now some of this she literally saw. She literally saw a light bulb. There's a light bulb image, and some of these things she was extrapolating, like '*how life is a big question*'. I have a lot of question marks in my book. I don't have this question written out, how life is a big question. She wasn't copying that. There's an image of a brain, but there's no image of my passion. How do you make an image?"
>
> "That was fascinating to me as an adult working with a then nine year old, how she was moving back and forth between the literal and her going from what she was seeing to what she was thinking about and what then she was wondering. Then, her response to what was missing was fascinating. She found that what was missing was '*what I like to do'*, the *'what else about me besides art?'*, which was referring to my going higher up your idea tree. The other thing that she said was missing was "firepower" When I asked her, "A, what do you mean by firepower?" I do remember, she was in third grade at the time. "Firepower is sort of like what happens with clay: it changes and transforms as we work with it. Clay starts as soft. Its first transformation is when the clay dries out, becoming brittle and fragile. The clay transforms again when it's fired and becomes hard and strong. Clay has the potential to transform again and again by adding layers of glazes. When artwork has firepower, it has transformed completely from the starting idea, the idea the artwork has layers. Firepower is also ***when the artwork crafted pops with strength and story***. You know, it's like readable."

This deep and impressive dialogic exchange was made possible by Caren Andrews sharing her accordion book with 9 year old A and asking her to interact with it. Afterwards the idea of "FirePower" became common parlance between A, Caren and A's classmates. Caren has "kept the term alive" as a metaphor in assessment : "The part about having an artwork crafted popping with strength and story can be both the visual elements that I'm looking for in assessing or self-assessment of artwork as well as the ideas behind the artwork. I think that that's why I continue to use the term "firepower" with that specific definition." We see Caren's centering of accordion book practice as an innovative paradigm of how the affordances of accordion books can add depth to teaching and learning.

6.10.3 Devika G: Things Talk to Me, Mapping Terrains, Getting Out of My Own Way

Devika is 20 years old and a scholarship student at the Srishti Institute of Art Design and Technology in Bangalore. She is also a youth facilitator with the placeARTs Youth Collective, an activist youth group that designs and implements community based art projects. Devika's primary languages are Telagu and Kannada and she has begun to learn English only recently. Devika began making accordion books as a middle school student of Arzu Mistry at the Drishya Kalika Kendra, a learning center for Urban Poor youth. She continues to make and keep her own accordion books today as a third year college student. She calls them *memory books* and keeps accordion books and flip books instead of regular note books. She has begun to seek inspiration by following artists she is intrigued by. Devika shares that her reasons for using accordion books goes beyond just an aesthetic choice. As a new English Language Learner, she finds it easier to share her ideas through a combination of words and drawings.

> "I started learning English very late. So I was not able to write a proper sentence or a phrase so it was easy for me to draw it. It was easy to take whatever sentence was in my head, and show it rather than talking about it. I do not feel more confident when I talk in English or write in English so I started using these journals and paper as my medium to share whatever was on my mind." (G, Devika, personal interview with Mistry, August 15, 2017).

Devika enjoys making books and has many different kinds of books that she makes and keeps but she speaks specifically about the versatility of accordion books. Besides being easy to make, Devika enjoys that accordion books because they can be small and transportable. She says "size matters but if you want to work big you can create a flap or extension, turn the book sideways and draw across multiple pages." In her other books she is restricted to size. "This book is like a mind map. You start with one topic with the bigger book and then you go to another topic where I have made a smaller book within the book. And then you just read about only that story. You do not get confused with everything else."

Devika also talks about how keeping accordion books makes her very happy. She says "I never get tired of working in these books. I am always trying to be creative in my books." Devika articulates that the aesthetic dimension of her book makes her look at other artist's work and builds her creative inquiry practice, She says "people come and ask me how did you make this and I go and ask people how did you make this? This way I am creating relationships .... my books make it easier for me to understand. All my books are for myself not some third person. I write in English and Kannada and just one word and one image helps me understand the whole knowledge of that day."

Devika shared one of her accordion books where she focused on observation. (Figure 6.7 left) "I have started drawing what I see and then I let my imagination take over ... this is a hut I was looking at but then I started colouring only one part and taking it into a tree. So adding my imagination to reality." These images still function as triggers to Devika about what she observed and also as metaphors for where her thinking grew from her observations. In Figure 6.7 (right) we see another page from a recent accordion book of Devika's. The page has some drawing, collage, natural objects and

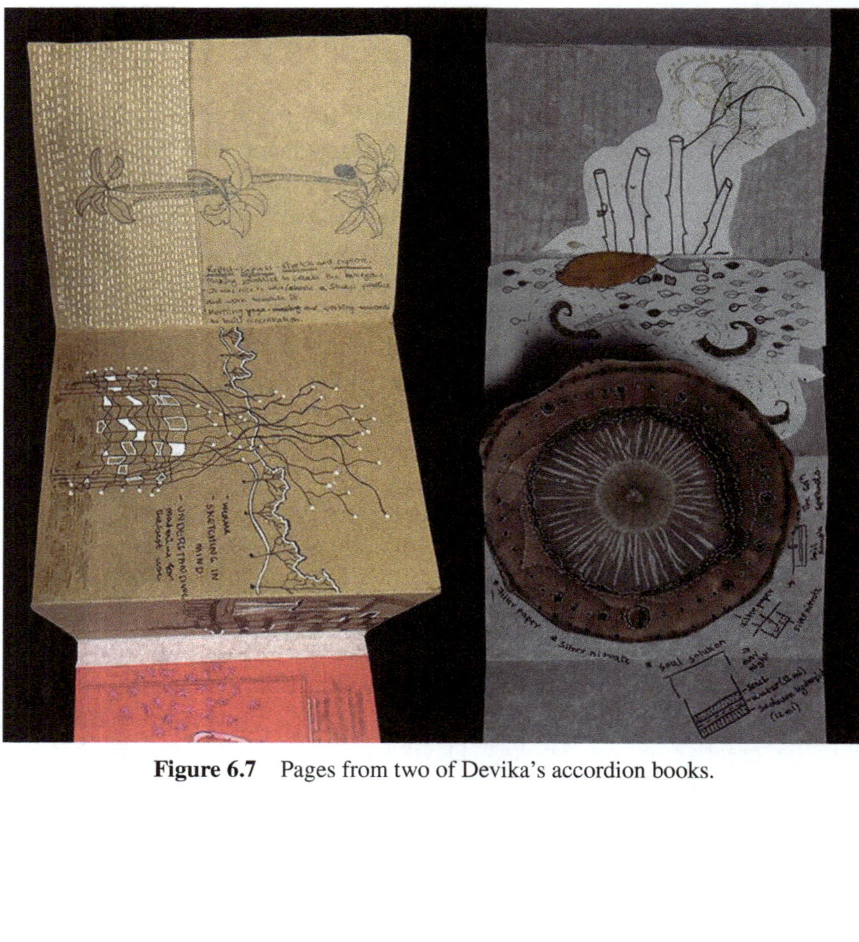

Figure 6.7 Pages from two of Devika's accordion books.

text. Devika shares a story about a night walk she had gone on that had huge impact on her. She saw multiple scorpions as she pointed her U.V flashlight at a pile of leaves. This 'things talk to me' moment was significant and emotional for Devika and she captures it here with the scorpion tails and leaves she has drawn. The circular image represents Devika's experiments with soil chromatography at an art/science workshop she attended. Below the circle she has text that documents the process, materials and directions. She has gone back into the image and layered it with white lines. This is the beginning of a map of Devika's experiments with wanting to understand a place in a deeper way. The chromatography in itself is a map of the soil.

While Mistry was looking at this page together with Devika, she expressed some frustrations about the process that has value to our discussion here. Devika was concerned that her accordion book looked too much like Mistry's

> "I don't want my books to look like yours. The thing is I have grown with you as a student and artist so it will happen whenever you grow with someone who is older than you you will take inspiration and use it in your work. But then even if you don't feel that you are copying them, others tell you. So many people said that you have a style of Arzu akka. I am glad but after a point I do not want it to happen. So I took inspiration from you not only in my books but in the way I talk, write or think about space design and then I mix it up and find my own way to start working towards it."

Devika expresses that she had been trying to break away and develop her own strategies in her books and she does not like the similarities between the look and feel of her books and her former teacher's. This conversation was important to have and one that Arzu has had with many teachers and students. She encourages everyone to copy ideas from each other and through the process make the ideas their own. Arzu responded to Devika by showing her the ways in which she was in fact innovating, branching off in her own individual ways, both in form and content. She also guided her to other artists she could take inspiration from. Like the dialogues between Caren Andrews and her student A, Mistry's conversations with Devika highlight the many ways accordion books can be a hub of teaching, learning and art-centered inquiry.

6.11 Some Challenges to Accordion Book Methodologies

Like any pedagogical methodology new to one, and especially a methodology that may go against the grain of past and current orthodoxy, accordion book practice may present challenges to those new to it:

Some learners and teachers not used to being invited to be the drivers of their own learning or to document and share their emergent processes (because this invitation has not traditionally been extended to them in their other learning environments) are reluctant, hesitant and sometimes unable to see the relevance of this kind of shift in learning/practice. This is due in large measure to learners' and teachers' general feeling of disempowerment often fostered by systemic compliance-based models of teaching and learning that do not encourage teacher or learner agency.

Students skilled at strategically "playing the school game" or delivering to the teacher precisely what is required to get the top grade, are not necessarily authentically engaging in accordion book practice, because they are not yet "doing it for their own reasons". Accordion book practice is most powerful when the inquiry, exploration, mapping, connection making, critical examination and revisiting is driven by the learner's own interests and is occurring in interstitial spaces between disciplines and students lives.

It can be challenging for teachers to stay consistent with accordion book practice, in other words to come back to it on a regular consistent basis. If teachers drop the ball, or fail to recursively create space for meaning making in and through accordion books in an ongoing way, they run the risk of sending a message to students that this practice is really not that important or valuable. For this practice to really take hold teachers must encourage the ongoing revisiting of accordion books-they must encourage a practice that will go beyond one's school day or school year, or even one's school life. In this way, accordion book practice can be processed and tracked over the course of years.

For some, a shift even occurs that one's accordion book practice/inquiry *is* the core and the classes one attends are just additions to one's core practice.

Many learners, both adult and young people, worry excessively about about their accordion books looking good. The anxiety of the blank page arises for many. Both Mistry and Elkin convey the message to their students that "copying is cool" and that "stealing is allowed and even encouraged". The goal is that learners eventually and organically get to a place of authentic and idiosyncratic skill – that they develop mastery that is self motivated and idiosyncratic versus one that is imposed upon them.

6.12 Conclusion

We have seen a great deal of evidence that accordion book practice supports the development of agency in teachers/learners/artists and is linked to an increased responsiveness to the world, a motivation to inquire and is a catalyst for developing ongoing inquiries and further practice. We've seen accordion book processes "jump start" the engine of self-directed learning in both adults and young people. As "sites of captured reflection" we've seen accordion books become a seed, a path, a constant place to revisit and move forward. We conclude, below, with some of the possibility spaces opening up for ourselves as a result of our personal continued engagement with accordion book practice. Our personal art-centered-research quests as teachers and artists have developed in parallel with the accordion book project as we practice keeping our own accordion books and work with others to keep their own.

There are different and often concurrent levels at which accordion book practice functions as art-centered research. We've witnessed examples of teachers and students separately but in parallel ways engaging in individual/personal lines of inquiry using accordion book practice and then engage in dialogue about each other's findings. Teachers' and students' inquiries thus have the potential to inform each other and suggest new individual and collaborative avenues for further research/learning. Examples shared here of Derek Fenner and Caren Andrews are just a small sampling of the work of many other educators and young people we have worked with. The cycles of noticing, gathering, sorting, analyzing, reflecting, coding, theorizing, developing new inquiries are a deeply personal and in-depth form of research. We see a great many possibilities in this type of arts-centered research. We will continue to publish and share exemplars of the innovations happening in accordion book practice.

Finally, although we have been engaged in the various processes outlined in this chapter for over 10 years now, as individual practitioners, collaboratively, and as educators of young people and other teachers, we feel that we are in many ways just beginning to explore the possibilities of accordion book practice. We are continually informed by the innovations of fellow teachers, artists and our younger students who have embraced accordion book practice and taken it in directions we would not have otherwise imagined.

Arzu Mistry has over the years been struggling with finding the fluid balance point and intersection of being an artist and teacher. Across her accordion books over the past many years, this challenge continues to

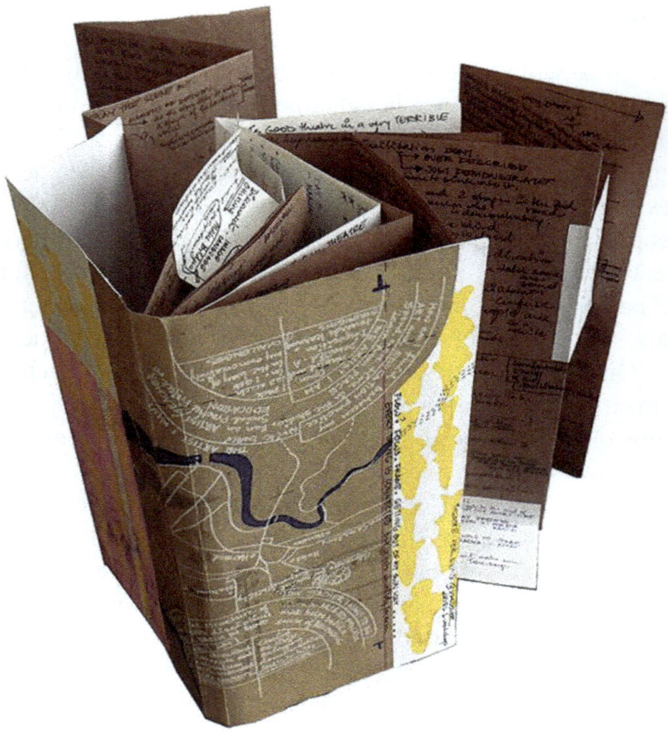

Figure 6.8 The Artist/Teacher I am.

surface in different ways and her inquiry still persists. She finds herself often replaying a conversation she had with her college professor who warned her that she could be a good artist or a good teacher but she would have to choose and she could not be both. Arzu has struggled to resist this dichotomy of artist or teacher for close to 20 years now. From 2014–2016 she engaged in an in-depth process of co-creating, designing, hand-binding, printing and publishing an artist book about the accordion book practice. Throughout the making of *Unfolding Practice; Reflections on Learning and Teaching* she grappled with how the process and product '*was* art' or 'was *about* art', as a lot of teaching is 'about' something else and rarely *is* the practice itself. There are points in the artist book where she feels successful in overcoming this dichotomy, and other points where she feels she failed to find that sweet spot. Therefore this inquiry is not resolved as yet and continues. As she grapples with this inquiry in her own practice, Mistry looks out at other practitioners like Joseph Beuys and Tania Bruguera for inspiration and simultaneously seeks opportunities to test the boundaries of this inquiry in her own teaching and artist practice. In a recent accordion book where Mistry has been inquiring into the artistry of facilitation and is engaging in the learning process of being a 'joker' in a Theatre for Living process for community dialogue, she reflects "Calm this analytical mind. It has stopped dancing. It only seems to be digging, poking, and unraveling. Just 'be' for a bit and create with fluidity not tightness." This inquiry exemplifies many aspects of the accordion book practice, from *getting lost* to *getting out of my own way* to *what would an artist do.* Mistry continues to use and reuse the strategies she and Elkin have developed through the accordion book practice in pursuit of an inquiry that is vital to her practice as a teacher and artist.

Todd Elkin, in his role as a high school art teacher, has been exploring various ways in which practices of teaching, accordion book processes, qualitative research, critical pedagogy and contemporary art strategies can merge. In addition, Elkin has been looking for ways that a fluidity between process and product can be achieved. A recent project, **Assessment as Dialogue** embodied these efforts.

Elkin had become increasingly dissatisfied with the fact that assessment in almost all public school classrooms travels in just one direction, from teacher to learner. In *Teachers As Cultural Workers*, Paulo Freire discusses the traditional practice of teachers' "Reading a class of students as though it were a text to be decoded" (Freire 2005) and then goes on to envision classrooms where students reciprocate, "observing the gestures, language . . . and behavior of teachers". Of course, students are already doing this all the time-however they are not typically invited to share the results of these "readings" with their teachers or with each other. With these ideas as starting

Figure 6.9 Assessment as Dialogue.

points, *Assessment as Dialogue* involved 30 High School students employing various analytic strategies and lenses to strike an assessment balance with an English Language Arts teacher whom they all shared. Utilizing accordion book strategies and other methods drawn from Critical Discourse Analysis, Relational Installation Art/Socially Engaged/Participatory Artistic practice and Qualitative Research and informed by the writings of Michel Foucault, Paulo Freire, Jeff Duncan-Andrade, bell hooks and other practitioners of Critical Pedagogy, students observed and unpacked verbal, nonverbal, written and digital communications produced by their teacher, all the while engaging in dialogue with him around their findings. The students used accordion books to collect, document, code and analyze their findings. A new qualitative research tool was thus invented specifically for this project, The Graphic Memo, consisting of each student's accordion book with all of its attendant gathering, sorting, coding, and analyzing strategies in play. Throughout, the teacher being assessed seized upon this opportunity to reflect upon his own practice. In addition, the students created an ongoing process-illuminating interactive art installation at a San Francisco contemporary art museum which also featured a series of public dialogues between stakeholders within their teaching/learning community and interested people from the wider Bay Area community. This Project began in classrooms at Washington High School in Fremont California and subsequently moved across the Bay where it became a dialogically driven installation at the 2014 **Bay Area Now** exhibit at the **Yerba Buena Center For The Arts** in San Francisco. This iterative art exhibit resembled a gallery sized accordion book, with the primary difference being that the students were continually focusing, reframing, highlighting, curating and re-presenting different aspects of their dialogic research findings on the gallery walls. For Elkin, Assessment As Dialogue, presented some promising ways forward in how accordion book practice can be situated within hybrid spaces of relevant transdisciplinary inquiries, teaching and learning and contemporary art practice.

In closing we would like to reiterate that accordion book practice has as much value for all learners be they teachers or students. We firmly believe that students and teachers need to develop and take ownership of a rigorous *practice*. This practice is not for school, or parents, or teachers or for a future job, this practice is for oneself. It is a practice one is passionate about and becomes the driver for continued learning. The philosophical base of accordion book practice as articulated in this chapter, fits seamlessly with the processes and strategies described. The microcosm is reflected in the macrocosm where an everyday practice is reflected in the very being and doing of the proactive and engaged artist and inquirer who is pursuing it. Olivia Gude in her talk on Art Education for Democratic Life says:

> "How does this engaged, aware person participate in a democratic society? First, the artistically engaged individual couples intense awareness with a strong sense of agency, a belief that he or she can shape the world. This belief in the average person's creative power lies at the root of any democratic society. As democratic citizens, we must believe that what we do affects the world around us, that what we do makes a difference." (p. 1)

Acknowledgements

Special thanks to Caren Andrews, Derek Fenner and Devika. G. for sharing their accordion book practice with us. A special mention and thanks to all the teachers and students who have taken this practice and made it their own.

References

Allen, F. (ed.) (2011). *Education: Documents of Contemporary Art.* Cambridge: MIT Press, 84–87.

Andrews, C. (2017). *Personal Interviewed by Todd Elkin Art 21*: www.art21.org, [accessed July 26, 2017].

Berger, J. (1972). *Ways of Seeing: Based On the BBC Television Series*. New York, NY: Penguin Books, 130–134.

Biesta, G., Priestley, M., and Robinson, S. (2015). The role of beliefs in teacher agency. *Teachers Teach.* 21, 624–640.

Bishop, C. (2012). *Artificial Hells: Participatory Art and the Politics of Spectatorship*. New York, NY: Verso.

Bishop, C. (ed.) (2006). *Education Documents of Contemporary Art.* Cambridge, MA: MIT Press.

Bishop, C. (Ed.). (2006). *Participation Documents of Contemporary Art.* Cambridge, MA: MIT Press.

Boal, A. (2002). *Games for Actors and Non-Actors,* 2nd Edn. New York, NY: Routledge.

Bourriaud, N. (1998). *Relational Aesthetics*. (Paris: Les Presse Du Reel), 113.

Calvert, L. (2016). *Moving from Compliance to Agency: What Teachers Need to Make Professional Learning Work.* Oxford, OH: Learning Forward and NCTAF.

Chan, P. (2014). *Selected Writings 2000–2014.* (Basel: Laurenz Foundation), 188–190.

Csikszentmihalyi, M. (1990). *Flow: the Psychology of Optimal Experience (Harper Perennial Modern Classics).* London: Harper Perennial Modern Classics.

Duckworth, E. (2006) *The Having of Wonderful Ideas.* New York, NY: Teachers College Press.

Dylan, B. (2005). *Chronicles Volume One.* New York, NY: Simon & Schuster.

Dweck, C. (2006) *Mindset the New Psychology of Success.* New York, NY: Ballantine Books.

Elgin, C. Z. (1999). *Considered Judgment (Princeton Paperbacks).* (Princeton, NJ: Princeton University Press), 146–169.

Emirbayer, M., and Mische, A. (1998). What is agency? *Am. J. Soc.* 103, 962–1023.

Eno, B., and Schmidt, P. (1975). *Oblique Strategies.* Available at: http://www.oblicard.com/

Fenner, D. (2017). *Personal Interview by Todd Elkin.* New York, NY: Art21, Inc.

Freire, P. (1970/1993). *Pedagogy of the Oppressed.* New York, NY: Continuum International Publishing Group.

Freire, P. (2005). *Teachers as Cultural Workers.* (Cambridge: Westview Press), 88.

Gardner, H. (2011). *Truth, Beauty, and Goodness Reframed: Educating for the Virtues in the Age of Truthiness and Twitter.* (New York, NY: Basic Books), 53.

Gude, O. (2009). *Art Education for Democratic Life 2009 Lowenfeld Lecture by Olivia Gude.* Available at: https://naea.digication.com/omg/Art_Education_for_Democratic_Life

Hara, K. (2015). *Ex-Formation.* Zurich: Lars Mller Publishers. 15–26.

Hetland, L., Winner, E., Veenema, S., and Sheridan, K. M. (2013). *Studio Thinking 2: the Real Benefits of Visual Arts Education.* (New York, NY: Teachers College Press), 58.

Horowitz, A. (2014). *On Looking: a Walker's Guide to the Art of Observation.* (New York, NY: Scribner), 2.

Kent, C. (1967–1968). *Some Rules for Teachers and Students.* Available at: https://www.brainpickings.org/2012/08/10/10-rules-for-students-and-teachers-john-cage-corita-kent/

Krysa, D. (2014). *Creative Block: Discover New Ideas, Advice and Projects From 50 Successful Artists*. (San Francisco: Chronicle Books), 22.

Mao, B. (1998). *An Incomplete Manifesto for Growth.* Available at: http://www.manifestoprojet.it/bruce-mau/

Marshall, J., & D'Adamo, K. (2011). Art practice as research in the classroom: a new paradigm in art education. *Art Educ.* 64, 12–18.

Mistry, A., & Elkin, T. (2016). *Unfolding Practice: Reflections on Learning and Teaching.* Kingston, NY: Women's Studio Workshop.

Narayanan, G. (2008). Keynote Address Symposium on Education & Technology in Schools. *Converg. Innov. Creat.*

Perkins, D. (2014). *Future Wise: Educating Our Children for a Changing World.* (San Francisco, CA: John Wiley & Sons), 23.

Perkins, D. (2009). *Making Learning Whole: How Seven Principles of Teaching Can Transform Education.* (Hoboken, NJ: Wiley) 203.

Pevzner, N., and Sen, S. (2010). *Preparing Ground: An Interview with Anuradha Mathur and Dilip da Cunha.* Available at: https://doi.org/10.22269/100629

Purves, T. (2014). *What We Want Is Free, Second Edition: Critical Exchanges in Recent Art,* 2nd Edn. New York, NY: New York Press.

Rinaldi, C. (2006). *In Dialogue with Reggio Emilia: Listening, Researching and Learning Contesting Early Childhood Series.* (London: Routledge), 25.

Schwartz, J. (2013/2014). *A Conversation with James Siena. Figure Ground.* Available at: http://figureground.org/a-conversation-with-james-siena/

Smith, P. (2016). *M Train.* New York, NY: Vintage.

Solnit, R. (2006). *A Field Guide to Getting Lost.* New York, NY: Penguin Books.

Turchi, P. (2007). *Maps of the Imagination: the Writer as Cartographer.* San Antonio, TX: Trinity University Press.

Wheatley, M. J. (2009). *Turning to One Another: Simple Conversations to Restore Hope to the Future*, 2nd Edn. San Francisco: Berrett-Koehler Publishers.

7

Practice-Based Reflections of Enabling Agency through Arts-Based Methodological Ir/Responsibility

Dina Zoe Belluigi

The School of Social Sciences, Education and Social Work,
Queens University, University Road, Belfast, BT7 1NN,
Northern Ireland

Abstract

Arts-based methods are well-placed to enable disruptions to normative positioning of researcher, respondent and subject. This chapter draws on the author's reflections of opening the research processes to the possibilities of methodological ir/responsibility. It focuses on a selection of mixed-method projects where a significant contribution to the *validity* of the empirical research emerged from the arts-based methods employed, including the use of journal writing, story-telling, metaphoric and visual imagery. The discussion is structured around the validity of the methods for the purposes of generating data to inform the evaluation of and research on that which is often difficult and elusive to analyse in higher education. A particular contribution of the chapter is the discussion of how the construction of research participants informed both the data generation processes, and the analytic approach to the texts they authored. An argument is made for the importance of establishing conditions which enable the possibilities of participants' agency.

7.1 Disrupting Positionality in Educational Research

Informed by the post-colonial, post-apartheid context in which most of my consultation and research has been situated; my concern has been to try to do justice to the subject of my research, while bearing witness to the incommensurability of diverse perspectives and experiences within the fraught terrain of higher education (HE). Although informed by the critical tradition of educational research, I have found myself increasingly discomforted by a propensity of researchers to unwittingly speak for, and possibly silence, those we represent (Roberts, 2007). As such, I have sought to create ways in which the boundaries of educational research conventions and practices, as with pedagogies of possibility (Giroux, 1988), can be permeated to enable more conducive conditions for the agency of my research participants. In attempt to acknowledge and open my practices as a researcher beyond such complicity, I have used the pages of this chapter to reflect on my some of my own explorations of what might be loosely considered ‘arts-based methods’.

When producing singular research reports, as we most often do as academics, the alterity of the individual account and ‘little narratives’ are often consumed because “the power-relation of subject and object reduces the world to categories and concepts” with the result that “the concept is privileged over the actuality it pre- rather than de-scribes” (Miles, 2006, p. 94). Constraining academic conventions extend from the technical, such as word-limits, to the conceptual and ideological, including the legitimacy of the dispassionate academic tone over affective, personal narratives; reliability privileged over validity; statistical rigour and generalisability over lived, contextualised experience. This is because education research has most dominantly been viewed in modernist terms, grounded in highly individualistic assumptions based on subject-centred reason and enlightenment ideals (Peters, 1995). The methodologies dominant in HE Studies for the most part continue this modernist drive, with a “will to certainty and clarity of vision” embodied in the narrative realism of its preferred writing style (Stronach & MacLure, 1997, p. 4).

Over time I found more challenge afforded to my own positionality by those research orientations which aim for an inversion (and subversion) of the traditional ontology-epistemology hierarchy, to a relationship between knowledge and ways of understanding the world which attempt to enact an ethical, liminal relationship between self and other. Heteronomy has been described as ‘a practical critique that takes the form of a possible transgression’ (Foucault, 1984, p. 45) of limits. The attempt is to subvert conventions of consoling certitude which impose regulations of ‘truth’, and to allow for recontextualizing ourselves with a sense of responsibility to imagine and represent differently (Bain, 1995). To resist the desire for closure, in this chapter I reflect on how arts-based methods have enabled me to mobilize meaning and explore significance *with* my participants. I believe that the inclusion of participants in various interpretative processes, and the emphasis placed on metaphor, storytelling and narrative, have further mobilised the methodological processes to rub against homogenising regulations and representations.

7.1.1 The Possibilities for Methodological Ir/Responsibility

My experience has been that arts-based methods enable opportunities and conditions to open up research conventions and methodological choices to participants during the research process, and, later for readers of the research which may not fit the received mould of 'responsible' research.

The anxiety, for correct compliance to ethical conduct in my initial ventures as a researcher, has increasingly become supplanted by the more important desire to find ways to put the principles I held dear into practice. Although that initial accent on 'warranted assertability' (Bleakley, 1999) – the sense that the study is of value and is a trustworthy representation that allows the reader access to my thinking, rationale, analysis, interpretations – was assured, it felt insufficient. Perhaps due to a context where the decolonisation of authority and positionality of the academic/researcher were being actively questioned, I increasingly moved towards including my participants to unearth, be critical, challenge and communicate my reflexivity on an epistemological level as a researcher. Such critical consideration of what frames my vision as a researcher, as 'epistemological reflexivity' (Hickman, 2008), involved being open about how my assumptions fed into the construction of knowledge generated within the report, which I tried to hold in balance with fulfilling my obligation to do justice to the subject being researched.

While I crossed my t's and dotted my i's, trying to be certain the process and product was 'correct', I felt that aspects of the research became stultified and reified, in particular for the actors involved. I began, without quite realising it at first, to desire methodological irresponsibility. And so I began to seek out those researchers and practices who looked for ways in which one may create possibilities and opportunities for transgression. Some of these practical opportunities included, explicitly inviting participants to shred the questions posed, to re-write or circle that which they felt was not representative; to annotate in the margins; to reject my summarized accounts where they were insufficient; and to tell their stories as and how they saw fit. In such moments, I described these disruptions to participants as a form of anti-authoritarian and playful graffiti. While the power dynamics between us cannot be entirely negated, such 'transgressive validity' (Lather, 1993) works to provide participants *the opportunity* to move from being passive respondents towards being active agents in quite pragmatic ways.

I sought ways in which to shift the conventional positions of power and meaning between me as a researcher and the person sharing insights – my 'respondent', as embedded within this term is a linear relationship. Informed by traditions of critical analysis, I wanted to commence from the experiences of those oppressed or those aspects repressed in order to understand the dynamics of structural power relations (Leonardo, 2004). It was important to disrupted received acceptance of being the one in power, or the one who knew more – in the hope that as authority would be shared, and both the processes of authoring and interpretation of stories would be more just. Over time, reciprocity, between meaning and power, researched and researcher, proved an important principle to transition beyond conventional normative objectification of the subject/object dualism which typify much educational research (Cook-Sather & Alter, 2011; Lather, 1991). The desire has been to find ways for active liminality, neither over-identification nor over-objectification. My approach was informed by grappling with Derrida's (1981) notion of ethical relationships.

The object of research cannot be closed because representation is subject to contingency and the historical moment of that reading – this acknowledgement enabled productive elements of self-doubt and scepticism in my thinking to translate into material possibilities of opening to participants in my research process, and to contest, in my representations. However, such openness is a difficult concept to practice in current educational research, with the conventions which valorise closure ('conclusions') and certainty ('findings') I thereby exerting power in its interpretations and constructing a consoling metaphysics of presence (Stronach & MacLure, 1997).

The validity of an uncertain methodological approach is echoed in recent studies on the negotiated space of the uncertain curriculum (Kalin & Barney, 2014; Wallin, 2008) in addition to the uncertainty in education in a supercomplex world (Barnett, 2000). Arts-based methods are often inclusive of idiosyncratic, pluralistic and individual contributions, and thus are characterised by such tentativeness and uncertainty (Stewart, 2008). Similarly, the pages in this chapter serve as sketches and interwoven reflections on my use of arts-based methods in education research. Although uncommon in the practice of HE researchers (Tight, 2004), I outline the methodological choices I have made within this chapter to be openly ideological about the development of my philosophical orientations to arts-based approaches.

This is to enable both my own reflexivity while in process, and to create openings for choice, interpretation and scepticism in your reading of this text.

7.2 Methodology

Acknowledging that the processes of research involve construction rather than passive discovery, in this section I discuss how I have attempted to be practically and morally careful of the ways in which the 'reality' of this chapter is constructed and the way I respond to the criteria with which it is judged by the editors and by you, my imagined reader. Although I cannot determine its reception, I see it as my authorial responsibility to take my intentionality and its potential consequences seriously. A sense of obligation and moral responsibility of an ethical imagination has, and continues to, weigh on me, perhaps due to 'historical melancholia' (Belluigi, 2001) of my generation of post-colonial artists and academics.

Over the past decade and a half, through some of the various research projects which I touch on in this chapter, I have wrestled with the identity politics of researcher, academic, teacher, artist, mother and 'white' woman, who until recently lived and worked in a so-called developing country in the global South. Added to this, is my pluralistic background in fine arts visual practice and in staff educational development. At the time of writing this chapter, I am negotiating how these identity threads and experiential knowledge(s) might fruitfully come to bare on international networks of HE Studies from a Northern Ireland location. In many senses I am operating with the privileged uncertainty of an 'émigré consciousness' (Said, 1993) which allows for a critical yet sensitive eye of politics, problematics and possibilities.

Excluding the practice-based research of my art making, my Master of Fine Art research began a search into understanding how to ethically relate to the other (both outside and within oneself); the difficulties of living as an artist under the weight of history; and the responsibilities of representation (Belluigi, 2001). Despite the wealth of this content, my methods of data generation remained firmly boxed within conventional academic approaches. A different lens through which to consider the human experience was developed through my Master of HE Studies, where I engaged with a more critical methodological orientation. Here I began playing with a dialogical relation between form and content (Belluigi, 2008a), where the form of my data generations methods diversified in response to my heteronomical orientation, and extended to include journal-writing and storytelling. I then began to advice and provide support for the exploration of alternative methods of data generation for teaching staff, when acting in the role of educational developer, and to model the use of such methods to those academics who were participants of my formal courses on teaching and learning. For instance, when looking for methods which recognised the significance of the affect on learning, I supported the design of maps, drawings, free writing and the integration of visual imagery.

Progressing through various research projects and evaluation approaches, including art making and practice-based research, I have found myself opening more profoundly to the interactional complexities of research. As discussed in 3.2, the possibilities for transgression and creativity came to fruition in my PhD research project, where I utilised various hybrid methods, including report and respond questionnaires, interviews and small group discussions; and visual narratives focus group discussions. In a current research project, I have continued such emergent data generation approaches, including metaphors and postcards.

Extending beyond the benevolence of the critical tradition are approaches falling under the umbrella of 'postmodernisms of resistance' which seek "to deconstruct modernism and resist the status quo' (Foster, 1985, pp. xi–xii) to open up potentially totalising narratives to difference, and so traces of the emancipatory intent of the Enlightenment were sceptically maintained. This nonfoundational tradition of research holds that, instead of metaphysical or epistemological bases, when a pursuit of knowledge has ethical implications, it should have an ethical basis and require justification (Smith, 2004). For this reason, you will find my reflections in this text often involves assertion, and sometimes backtracking – as I reconsider the decisions made and reflect on participants' feedback and experiences of such methods. When it comes to analysis, instead of being dictated to or grounded by a dominant framework of understanding or operating within its context of expectations and values, I see the analytical tools I utilize as 'openings' resisting the closure and surety of generalizable conclusions. As I have often explored problems rather than prescribed solutions, such an emergent approach has stimulated thought and generated problems around and about the 'the field of disputed meaning' (Stronach & MacLure, 1997, p. 113).

As this chapter focuses on how arts-based approaches have enabled participants' agency, I will focus here on a discussion of the analytical construction of structure, culture and agency in my attempt to grapple with that which is espoused in HE, and the significance of that which is experienced. To do so, I most often utilized critical discourse analysis when looking at representation, asserting a distinction between discourse and narrative.

The understanding I have held is that discourses do not determine identity but provide the conditions within which they are negotiated (Foucault, 1979, 1980). The person is understood as a psycho-social subject (Davies & Hare, 1990) shaped by and shaping him/herself through shifting identifications with the various discursive positions in which s/he is situated.

Of the many distinct orientations to narrative in research, there are two which I find helpful to distinguish: the self as constructed or revealed by the representations, or the self as concealed by them (Sclater, 2003, p. 318). In the former, narratives are central to identify-formation, through which significance is ascribed to experience and the self is constituted. Researchers in this intentionalist model analyze narratives as stories of an individuals' autobiography which enables privileged access to the author's view of him/herself. This understanding of narrative is underpinned by a metaphysics of presence, where an authentic, autonomous self creates a representation as a private object which is the most correct or close version of that person's meanings (Parker, 1997). A more critical view is that narrative should not be taken on face value – the researcher should be sceptical, partly because of aspects of the self that are beyond the bounds of conscious discourse, and partly acknowledging that the 'defended subject' may unconsciously or consciously alter or manipulate the stories s/he tells to defend against the 'real' self (Holloway & Jefferson, 2000). Both approaches recognise that there are complex connections between narrative and identity, and thus the study of narrative is epistemological (Stewart, 2008).

There are elements and layers of subjectivity intrinsic to the narrative and the act of narration itself, that are important to consider when constructing the narrative 'self' and analyzing the narrative. These include the speaking subject (the 'actual' person) who invites or addresses 'the subject' of the speech or text (the imagined reader), and creates a 'narrating subject' (the narrator), construct the subject of narration (the character) to speak about the narrated subject (the thing to which the narrator refers but cannot get there because of language – the signified) (Sclater, 2003). In my analytical processes I have used these differentiations to acknowledge the agency of my participants, in how they choose to describe their experiences and the 'self' they construct in the text, particularly in how they respond to, resist, manipulate, or collude with larger discourses, my own presence as researcher, and their imagined audience. This notion, of the many subjectivities in narrative, has freed me from the notion that the layers will correspond. However, I have not found extreme anti-intentionalist constructions, of narrative as pure fiction, productive.

Stories both help us understand the social and cultural context within which the person is situated, and reveal idiosyncratic characteristics, in a way that echoes the neither/nor of ethical self/other relations to which Derrida (1981) refers. In the space between neither/nor is the 'always-becoming' which is psycho-social and involves ongoing 'identity work' when negotiating the politics of belonging as a human agent. Narrative is a dynamic practice of active, intentional and embodied agents which is simultaneously individual and relational to social, cultural and interpersonal locations. Narrative acts as a 'potential' or 'transitional' space where the self is created or transformed in relationship with others and within the matrices of culture (Winnicott, 1971).

These narratives are located within or contribute to the larger discourses, which as artefacts of culture, can be 'read' for both meaning and significance. Informed by Foucault and Critical Theory, I have utilised critical discourse analysis to explore rhetorical power-plays in HE which de/legitimize narratives, regulate meaning and determine criteria which are used for judgment. While analysing, I seek to make explicit discourses, often within participants' teaching and learning interactions or larger aspects of institutional culture, that were otherwise implicit or invisible, and thereby more powerful, with the intention of exploring the political, social, cognitive and affective *significance* of such discourses. Cultivating a sensitivity to or awareness of discourses within research relationships is a means of consciousness-raising in the hope of demystifying their 'taken for granted' nature within narratives and on the power of their positioning of subjectivity.

Against conventional representations of the individual researcher as 'ideal knower' detached from history, affiliation or cultural bias, for my participants and my readers. In this section, I have sought to make visible the factors and practices which shape my choices of arts-based methods as a researcher. In the sections following, I discuss the challenges and rewards of the arts-based methods with which I have worked, the analytical approaches I have found most fruitful, and some of the dominant limitations, concerns and cautions of such methods.

7.3 Practice-Based Reflections on the Purposive Validity of Arts-Based Methods

I have shaped this discussion of arts-based methods in relation to the central purposes of the projects within which they were utilised: staff educational development, institutional evaluation research, and research on assessment in HE. Within my educational development role, my influence was positioned as that of an advisor or informed critical friend. However, I was able to develop and thereby model more divergent approaches within the evaluation practices adopted in my teaching of formal HE Studies courses, in addition to utilizing arts-based methods in my research involving student and staff participants as I discuss in the next section.

7.3.1 Creating the Conditions to 'Listen' to Participants' Experiences in and of HE for Evaluation Purposes

The institution, in which my educational development work was situated, had adopted a formative approach to evaluation which enabled autonomy in how academic staff determined their own evaluation agendas, methods and approaches. While some used this latitude to avoid accountability, others explored methods which generated rich insights into their practice and their students' lived experiences. Of importance in that rapidly transforming and contested HE context was to develop evaluation methods towards transforming curricula responsively; to enable student voice and increase student ownership; to create opportunities for inherent teaching and learning practices to be challenged (Belluigi, 2013b). I curated an anthology of selected case studies of such approaches (Belluigi, 2016). In this section, I focus particularly on the arts-based methods used by such staff whose evaluation practices I had directly informed, through the emphasis I asserted during workshops, formal qualifications, and individual consultations. One of the currents within the institutional milieu was to foster a culture appreciative of multiculturalism within the context of inherited troubled history and inequality (Jansen, 2008). Informed by the critical tradition in adult education, the formal courses broadened the focus on the individual teacher and narrow understandings of curriculum design to exploring the contexts, circumstances and conditions most likely to encourage and maintain student involvement and investment. The underpinning impetus of such attempts were to engage with the experiences and desires of members of groups who had previously or who were suffering forms of oppression within the politics of belonging in the teaching-learning space or institutional culture; to uncover mechanisms of domination; and to support struggles and innovations against inequality.

Learning engagement, as inclusive of cognitive, affective, connotative and relational aspects of learning, should extended the horizon of how data is collected and feedback generated. Due to an awareness of the nuances involved in accessing affective aspects, such as experiences of alienation and engagement (Mann, 2001), in my interactions with staff I encouraged that the generation of such participant insights be approached in more unconventional and exploratory ways than the conventions of summative measurement often permits. As such, free writing activities, metaphor, word descriptors, journal writing and imagery, amongst others, were introduced in the content of formal courses in addition to being engaged with experientially as data generation processes by the academics who were participants of my courses. Such experiential knowledge gave participants a tangible sense of how enabling 'voice' and legitimising lived experience may empower participants as agents within teaching and learning cultures and structures, and allow for learning experiences that are reciprocal. By recognising the importance of the student investing and exploring his/her personal stance in the learning process, a number of academics in turn created opportunities in their own curricula for increased student ownership, responsibility and co-production in teaching-learning processes (Belluigi, 2016).

In my formal courses, I actively facilitated such a shift in conceptualising evaluation instruments as integral to teaching-learning interactions. According to Shor (Brookfield, 1995, p. 93), the 'first responsibility of critical reflective teachers is to research what students know, speak, experience, and feel, as starting points from which an empowering curriculum is developed'. The data generated then allows academics the possibility of comprehending the complexity of learning experiences from the 'other' side, to challenge how our practice might create environments and activities more conducive to encouraging engaged and committed experiences for participants in their contexts (Belluigi, 2008b). While qualitative responses are most valid for these purposes, I have often suggested that quantitative responses are often generated or deduced too so as to ensure the strategic impact value for external stakeholders, or as an indicator of extreme responses requiring further probing. Of import, is that such instruments and approaches were designed in a contextually responsive manner (Nygaard & Belluigi, 2011).

The rich validity of literal and metaphoric imagery was demonstrated to academics while participating in my course evaluation processes or in my own research. In response, a number of my peers utilised visual aspects in their data generation methods for evaluation purposes. For instance, the inclusion of such elements as emoticons (Van der Poel, 2016) and the construction of African American quilts (Seddon, 2014) are not conventional in academic research (Ptaszynski, Rzepka, Araki, & Momouchi, 2011). Whilst the utilisation of such visual elements has varied in terms of sophistication, it is important to bear in mind that the purpose has been the process they elicit and conditions they create, rather than the mastery of form or their slickness as a product.

The writing of narrative has long been recognised in educational development as a powerful way for participants to express their experiences, to engage in a process that may enable them to reflect on their learning journey. I encouraged my academic peers to be cognisant that for narrative to be more than a response, it should be given the space to operate as a story. Stories require listening, sharing understandings, empathy, rather than measurement. Because absolute, definite conclusions as 'evidence' cannot be drawn from responses to such ambiguous stimuli, there are those who are critical and dubious of such methods, particularly those concerned with summative evaluation for tenure, promotion or QA purposes rather than development or enhancement. My emphasis has been to ensure (and gather evidence to support) the *validity* of the instrument in terms of helping the students, and in turn academics, to engage meaningfully with the conceptual criteria of the course and with students' experiences and engagement with their teaching and learning culture, rather than overemphasizing claims of the reliability or objectivity of the instrument. In fact, such stories may act as Lyotardian 'little narratives' where both the act of telling and the implicit pragmatics of narrative transmission function to displace the scientific claims of narrative realism. Drawing from such stories adds an inbuilt mechanism to prevent claims of absolute certainty about the quality of the course or teaching, but to rather keep active ongoing dialogue which,

> *engages the student not simply as an active rather than passive 'receiver' of knowledge, but rather as an active creator of knowledge with the teacher (Grundy, 1987, p. 101).*

Particularly in a context where machinations of domination and prejudice are both overt and covert, insights from those who are less powerful are the most valid catalysts for informed rupture of teaching, learning and assessment practices which replicate hidden curricula.

7.3.2 Creating the Conditions for Participants to Author their Stories of HE

Similar to what I encouraged in my courses and educational development role, I have actively sought to develop my own hybrid and contextualised research approaches. In this section, I discuss the arts-based methods utilised in three research projects as illustrative of my approach. To do so, I outline the research subject, the data generation instruments utilised, focusing in particular on the arts-based approaches.

The first is a project which explored the disjunctions between the espoused and practiced curriculum of a creative arts discipline by excavating the formative assessment method, the 'Critique' (Belluigi, 2008a). When it came to deciding sources and methods of data collection, I aimed to gather data positioned in gaps between theories espoused and in-use, and which would allow insights the significance for student learning. Multiple sources and methods were required to explore both declared aims of the curriculum and the underlying, non-observable processes of teaching-learning interactions. I drew data from academic literature; various 'texts' produced by the institution; interviews, discussions and questionnaires with participating academic staff; participating students' journal and stories; and my observations of the assessments events. Using critical discourse analysis, data collected and generated about and during the event of the formative assessment was analysed to unlock the unexamined assumptions and beliefs of the teachers (Belluigi, 2009), and the experiences and approaches of students. What emerged was that the dominant discourses in the case studied constructed a negative dialectic of the artist-student that denied student agency and authorial responsibility (Belluigi, 2011). Students experienced this as alienating, to the extent that to preserve their sense of self, they adopted surface and strategic approaches to learning.

The richest data was generated through the arts-based methods of journal writing and third-person storytelling. I designed a hardcopy daily journal in which students were asked to write/draw/express their experiences of their learning process before and after an assessment event (see Figure 7.1). Feedback on the design had been elicited from teachers of multi-disciplinary backgrounds, including educational development, psychology and fine art, informing refinements. Visual elements of the journals differed according to the subjects the students' studied, to create a more contextual aesthetic and identity for each participant. At that period of time, hardcopy journals were more inclusive of variations in participants' socio-economic backgrounds.

Following the event, the student participants were invited to a synchronous meeting to write stories of their experience over that given period of time. Participants drew from the reflections-in-process they had recorded in their journals to construct their own stories, and in this way actively directed the first step of the analytical process. I suggested they write about their experiences in the third-person, and some created names for themselves. To ask such disclosure and enable their stories to grow, required that I temporarily

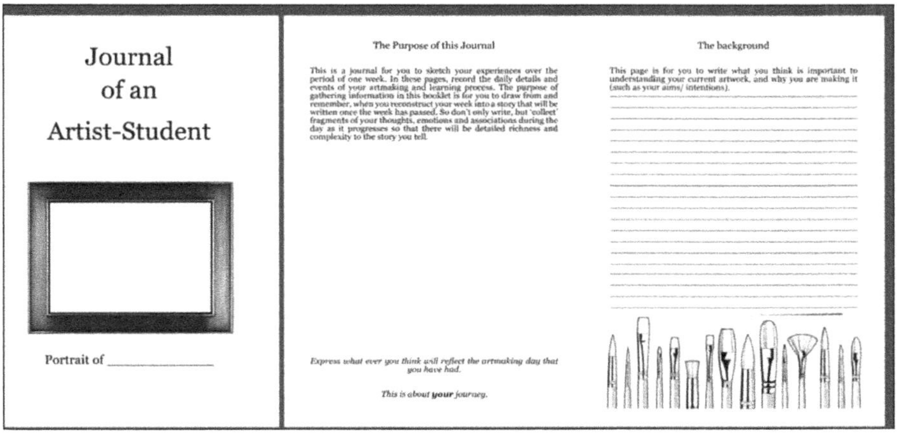

Figure 7.1 The first three pages of a journal designed for this project.

forgo the role of researcher for an appreciative listener who exhibits sincere interest (Silverman, 2007) in all the difficulties and details. In this shift, such approaches seemed more ethical in avoiding objectifying the participant. Derrida (1981) argues that by elevating its own record, the group in power de-stabilizes and threatens to extinguish the value of individual memory. By privileging such acts of representation, I explicitly acknowledge postmodern uncertainties about what constitutes an adequate signifier of social reality.

As discussed earlier in this chapter, a fruitful way to facilitate participants' reflexivity to unearth the assumptions which underlie practices is to approach them as if foreign, as discussed in 7.2. This notion was communicated when I suggested students compose stories in the third person to create aesthetic and contemplative distance. Whilst I would not claim that this project had the potential of 'consciousness raising' of the Women's Movement or 'conscientisation' of Freire (1972), I did intend the process to increase participants' awareness of how,

> *hidden below the surface narrative of stories are the assumptions, models, expectations and beliefs that guide people's decisions and behaviours... stories about real or imagined situations tend to capture these underlying assumptions (Silverman, 2007, pp. 34–41).*

For instance, the research process a number of participants came to realise they had developed a 'false self' to survive the assessment practices (Winnicott, 1971). The excerpts below indicated that they had learnt to approach the formative event strategically rather than as an opportunity for honest disclosure.

> *It is Wednesday, the day before Beatrice's long-dreaded crit... Is she to invent a whole new string of fiction that justifies her work or is she to re-tell her previous concoction (Beatrice's story).*
>
> *Personally, I'm learning the fine art (pun intended) of crits. I treat it like a performance, or a presentation, even a lecture. I spend a lot of time scripting what I am going to say, which is great because I can just read the script in the crit and not look at anyone's face (Penny's email).*

What such data enabled me to access were the contributions of the affect to how the student-instructor relationship was constructed. Such studies widen the focus beyond the product of the learning engagement, to recognise 'what they *experience* while a student: the life of a scholar in a community practicing its discipline' (Parker, 2003, p. 539). Believing in the importance of experience and the process of students developing as critical beings, I drew from Mann's (2001) seven perspectives of alienation to analyze the student data. Alienation in HE is not necessarily inevitable, however critical examinations of these conditions is necessary to inform radical changes to educational interactions.

As I shifted from a strongly critical orientation to research, towards opening myself up to heteronomical possibilities, I began to find ways in which instruments not only generated but also disseminated information. This was to progress participation to action – asking participations to further interpret, extend, supplement or problematize my interpretations. I enabled this shift explicitly in one particular research project, concerned with how differing interpretative approaches play out within referential frameworks in teaching, learning and assessment interactions in HE (Belluigi, 2017a). As with the project discussed above, I collected data from course documentation and generated rich data utilising a variety of hybrid methods, including observations of assessments, questionnaires and interviews with staff. However, at various points during such researcher-participant interactions, I explicitly and purposefully created possibilities for reciprocality, transgression and challenge of interpretations, as I discussed in 1.2.

In addition to data generated from students' responses to questionnaires, I further developed an arts-based method which had been designed specifically to generate data on student experiences for evaluation purposes

(Meistre & Belluigi, 2010) in a bid to extend the boundaries of the potential of imagery to explore the dark spaces of HE (Bengtsen & Barnett, 2017). Participating students constructed visual stories which they then discussed in small group discussions, facilitated by myself as a person external to the teaching-learning relationship. The instrument engaged students with mimetic activities as they played with their relationship and the meaning created in their stories; which, in turn, would better enable the course coordinator's reception of the particularity and diversity of each story. This was in the tradition of art making processes where

> *one is neither exclusively subjectifyingly inside one's own creative experience, nor objectifyingly looking in from outside the field or territory of work... on the boundary, wrestling relationally with the various conditions, inner and outer, practical and theoretical, creative and imitative, biographical and analytical (Dallow, 2003, p. 61).*

As this method proved particularly powerful in the richness of data elicited and the positive reception to the process by participants involved, I developed it further and utilised it considerably for research purposes. In what I have since called 'visual narrative focus group interviews', participants author their own stories using found images and text, which they then explain in small group discussions with their peers. These insights were triangulated with data from other methods and sources in relation to the optimal conditions for creativity in this domain (Belluigi, 2013a). Schema of the environment, relationships and curricula were then sketched, indicating the significance of interpretative approaches on students' emotional, critical and reflective engagement with themselves and the process and product of their learning (Belluigi, 2017b).

Because imagery has been used in healing and psychotherapy since before antiquity, data generation methods which utilize images are often informed by psychoanalytic approaches (Cabrera & Guarln, 2012; Prosser & Prosser, 1998). Such use of imagery is 'based on the principle that change of emotional and physical symptoms could be achieved by effecting a change in the imagination' (Edwards, 2011, p. 11). The contemporary psychoanalytic term 'imagery rescripting' encompasses a range of methods which utilize imagery to both assess and address a person's underlying emotions and meaning-making. Images are seen to emerge within consciousness and lie behind emotions acting as gateways to surface deep-seated issues and concerns evoked through experiences.

> *[The] image can, when properly understood, foster a deeper sense of the underlying meaning that [an interaction] holds for my sense of self within this particular sociocultural context (Dirkx, 2001, pp. 65–66).*

The cultural theorist Roland Barthes (1984) made much of the division between those images which contained informational and aesthetic value (a *studium*) and those where a shock, thrill or emotion is elicited (a *punctum*). The latter has the potential to activate the reader who is then drawn beyond that which is easily readable or received to a second level of meaning, as a *punctum* triggers 'a succession of personal memories and unconscious associations, many of which are indescribable by the individual' (Cronin, 1989, p. 72). While it can be argued that such a dichotomous psychoanalytic conception imposes artificial separations between the 'public' or overt message, and 'private' or personal interpretation, a constructionist understanding of the social nature of meaning-making can be applied to the participant's relationship with imagery. More textualist approaches to imagery acknowledge the mimetic moments which transcend 'the non-conceptual affinity of a subjective creation with its objective and unposited other' (Adorno, 2004, p. 80), enabling unintelligible and mysterious aspects of the world and 'the other' to emerge (Gebauer & Wulf, 1995). Advocates of projective testing have utilised purposefully ambiguous imagery so that the reader projects his/her interpretation to reflect their feelings, experiences, prior conditioning, thought processes et cetera (Kaplan & Saccuzzo, 2012). Informed by such notions, a postgraduate student of mine recently probed the identity positioning of first generation students (Alcock, 2017) by asking them to constructed photographs which represented them 'at home' and 'on campus'.

While most arts-based research methods are similarly informed by concerns with what the imagery conceals or reveals of unconscious impulses (Edwards, 2007; Shorr, 1983), visual pedagogy that is informed by critical theory, considers *the tactics of reading and writing* in the construction of visual narratives (Rifà-Valls, 2011).

> *A fact of primary social importance is that the photograph is a place of work, a structured and structuring space within which the reader deploys, and is deployed by, what codes he or she is familiar with in order to make sense (Burgin, 1982, p. 153).*

The design of the visual narrative instrument in my study was informed both by imaginative constructions and by projective psychoanalytic approaches, purposefully utilising ambiguous photographic imagery, so that the person could project and re-imagine him/herself into the situation to create a story. Participants construct a sequence of images, chosen from an image bank of ambiguous images devoid of human subjects, alongside their own written captions, in response to a posed statement or phrase to create a story. The found images were selected from an art archive (Meistre, 1998) which had been utilised used in a cross-disciplinary collaborative project between an artist and psychologist (Meistre & Knoetze, 2005), to both reference and upset the format of psychological projective tests. A sense of play and experimentation inverted the conventions of power, placing interpretative agency of these 'stories' with the participant.

Once the visual narratives were constructed, the participant shared it with his/her peers and me in a small group discussion, discussing the choice of images and text as well as the experiences they intended to evoke. This discussion of the participants' actual intentionality was the first act of interpretation of the visual narratives, which although organized by the intellect, took its impetus and meaning from the affect (Shorr, 1983). I asked further probing questions to comprehend the significance of such experiences, and to ascertain whether and in what ways such experiences extended to others. The discussions were audio recorded, transcribed and then after analyses, sent to the participants for further discussion (see Figures 7.2 and 7.3). These processes actively engaged participants with the 'polysemic' character of imagery, where each new discourse situation appropriates and generates differing sets of meaning.

In a project exploring the reception of equity-agenda staff development programmes (Belluigi & Thondhlana, 2016), postcards were utilized to probe academic staff members' experiences of institutional culture. As with the other projects discussed above, this was not the sole instrument of data generation. It was incorporated at an important juncture in the research process – at the closure of small group discussions in which we presented for further deliberation our interpretations of those participants' responses to a seven-page online questionnaire. The participants were handed an envelope of pseudo postcards, each printed with a metaphor which had emerged in the questionnaire responses: 'talk show'; 'alien space'; 'training the dog'; 'elastic skin'; 'poster child'; 'put into a pot'; 'gatekeeping'; 'window-dressing'. In this method, we capitalized on how metaphor accesses a different way of thinking to conventional research methods, wherein participants contrast, negotiate and manipulate imagery rather than rely on the linear, logical structure of language.

1. When a work I was invested in was assessed.

Fran presented a visual narrative to her peer group that showed a number of images cut up and reassembled with her drawing over them. Fran described how initially her feelings of having work she was invested in be assessed, were "dark" with "this overwhelming kind of dread". This was due to her fear and sense of exposure and vulnerability that work she had laboured over might be rejected. Her engagement with the assessment event felt like she was being made to attend or create "my funeral" with a sense of inevitability as if she was "headed straight for like a plane crash". She summed up this emotional state before assessments as "that kind of funk of it's all going to go wrong". Fran explained though that many times these fear turned out to be unfounded, where the growing anxiety lead not to catharsis but "a kind of anti-climax" in the mundane responses ("just a couple of pizzas") that is almost disappointing, unsettling but exhausting.

Figure 7.2 A reproduction of Fran's visual narrative with my analysis including excerpts from the audio recording of her description which she articulated to her peers in the focus group discussion.

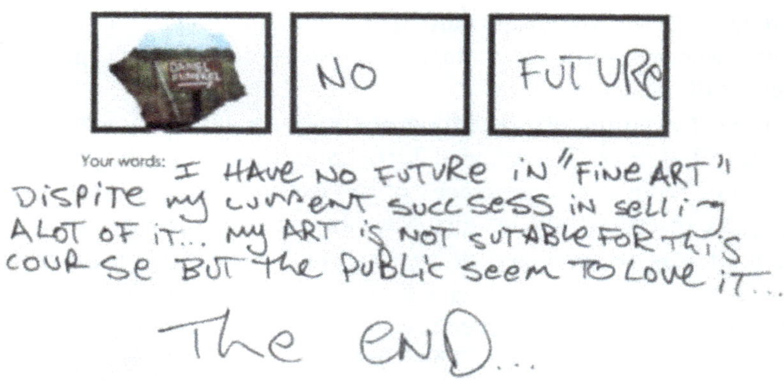

Joe's last narrative was about the sense that he felt at that point, 3 months from completing his fine art studies, that he would not be continuing engagement in fine art practice. Joe chose an image with a sign for a funeral, indicating a death or loss had occurred, and wrote the words "no" in one block and "future" in another, with the text beneath "I have no future in "fine art" despite my current success in selling a lot of it... my art is not suitable for this course but the public seem to love it.... The end....". There is a strong sense of an experienced conflict of external in/validation, in the

Figure 7.3 A reproduction of Joe's visual narrative with my analysis.

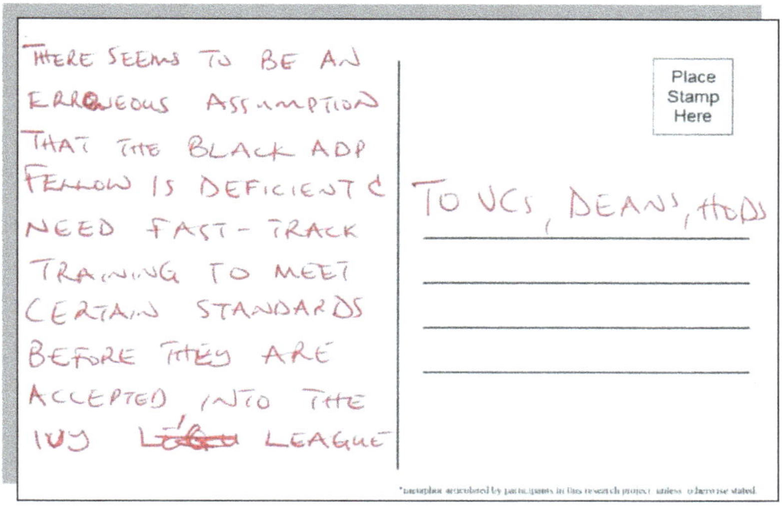

Figure 7.4 A postcard in response to the metaphor 'training the dog' addressed to those in the management structure of the institution.

The metaphors, in this project, acted both as emergent products of the initial research process and a way to deepen our relational processes with the participants and each other. We were most interested in the potential of 'the contingent, multiple and intertwined processes of visualising metaphors' to open up and deterritorialise identity (Boulton, Grauer, & Irwin, 2017, p. 210). This was integrally related to the content of the research subject, which was about the subtleties of personal and group-based identity and diversity, and its relation to perception, experience and performance in the evaluation of academic staff.

The information on the envelope indicated to participants the main reason for our inclusion of this method:

> *Choose postcards.*
>
> *Write on the back.*
>
> *Address it to whomever you would most like to 'hear' what you have to say*.*
>
> **This will help us decide the intended audience of the papers.*

By selecting their imagined target audience, the participants participated in deciding whom the readership of the dissemination of the findings should be.

Particularly with that research project, which considered the impact of societal asymmetries in the assessment of individual's performance on affirmative action programmes, this act positioned participants powerfully. Our decision was formulated as a response to the debate around the appropriateness of choices of readership in the politics of representation is pressing and pertinent (Leonardo, 2004).

With varying levels of intensity required from participants and preparation from the researchers, these three projects indicate a concern for utilising arts-based methods to explore the parameters of the authorship of the participants and the readership of the findings.

7.3.3 Concerns, Limitations and Improvement

There are those who raise questions about the reliability of arts-based methods which have a nonlinear relationship of representational narrative to the real (Chappell, Rhodes, Solomon, Tennant, & Yates, 2003). While some question the reliability of such methods and the currency of alternate approaches to research in our neoliberalist times, I have found that such methods are valid as they enable engagement with that which is tacit, nuanced and difficult to measure in educational practices. As I have discussed in this chapter, I have found them useful to explore the hidden curriculum, the gaps between what is espoused, the theory-in-use and what is practiced in teaching and learning, in addition to what is experienced by students. To combat perceptions of partiality in such methods, the triangulation of methods and sources have served to ensure warranted assertability by demonstrating that dependable conclusions are being drawn. I have found 'report-and-respond' approaches (Stronach & MacLure, 1997), in particular, ensures rigor through their transparent and inclusive role of generating, disseminating, supplementing and problematizing data and interpretations of the subject being studied. For instance, in the last two projects discussed, the initial questionnaires were composed of statements explicitly based on notions from relevant hyperlinked published literature, with the intention of eliciting participants' informed and considered responses.

As they are creative in and of themselves, methods which appreciate the critical role of the visual (Sullivan, 2007), such as arts-based strategies (Barone & Eisner, 1997; Diamond & Mullen, 1999) reinforce the interpretativist notion that research creates rather than discovers. This is because imagery, whether literal or metaphoric, can be 'compared with a complex sentence [rather] than a single word. Its meanings are multiple, concrete and, most important, constructed' (Tagg, 1988, p. 187). In their form, such methods embody the notion that 'the self' is constructed rather than discovered within structures such as education. Arts-based approaches can encourage participants to experiment and reflect by 'playing with, trying out, discarding identity, purpose, shape' (Parker, 2003, p. 541), as they actively engage with creating, shaping and interpreting their responses. Such approaches trouble the binary of constructs of cognition and emotion in binary opposition (Besnier, 1990) rather than positioning the affect as integral to ways of knowing the relationships between the self and the broader social world. The nuanced insights yielded by such innovation warrants the emphasis on process (Prosser & Schwartz, 1998). Arts-based methods hold the potential to provide more engagement with "deeper understanding of the emotional, affective, and spiritual dimensions that are often associated with profoundly meaningful experiences in adult learning" (Dirkx, 2001, p. 70). Because of the emotional effect of the metaphor, story or found image, there is more chance that it will break those sophisticated cognitive resistances to and conscious censorship of inter-relational dialogue and verbal and written transactions (Shorr, 1983) which typify conventional research approaches.

In addition to their validity for generating rich data, arts-based methods create conditions for powerful transformational learning experiences. They jolt the discursive familiarity of educational practices that are the focus of such research, disrobing their "mythical immediacy" (Buck-Morss, 1997, p. x). Such disruption serves to force both participants, and in turn readers, to look again or more slowly and carefully, at that which is often taken for granted by each perspective (MacLure, 2003). This is an important consideration for educational development, where building the capacity for reflexivity of teaching staff and students who may carry with them troubled histories or reproduce problematic traditions of adult education which replicate inequalities. One way to unearth the assumptions that underlie practice and experience, as well as how one constructs oneself in relation to discourses, is to approach them as if foreign. My understanding of such processes of making the familiar strange are heavily influenced by the argument that the experience of repressed strangeness or the uncanny is central to the enlargement of political imagination (Kristeva, 1991).

Some of the challenges of such instruments were to create safe-enough space for participants to share their experiences, and to 'buy in' to such alternative processes. Whether for evaluation or research purposes, by ensuring that the processes were low stakes, were informed by transparent ethical principles, and where I demonstrated a sincere stance of listening, created a climate encouraging of trust, play and transgression. However, removing social pressures and credit-bearing incentives definitely impacted on the participation rates, as indicated in the response rate of 11 of 40 invited students in the storytelling project and 26 of 81 invited students in the visual narratives project discussed in this section. When utilised as a way of formatively evaluating curricula, and thus with structured time for such interactions within contact time, participation rates have been far higher.

Another factor is that the quality of what participants 'produce' may differ in the initial interactions, particularly for those participants who start off by feeling intimidated by their abilities to express themselves creatively. With visual narratives, it important to build trust in the process, and to emphasise that mastery of the final product is not the point, but rather that participants understand that the visual aspects are only a part of their expression. I have found that responses have varied in effort and approach, with the process mostly appreciated as playful and engaging. Across the eight focus group interviews of the visual narratives project, for instance, only two out of twenty-six participants articulated concern that the images complicated the clarity of what they hoped to communicate. They exercised their agency to engage with the narrative as they felt best representative. As there were opportunities to describe the narratives within the focus groups, and to further add to my analysis later on in the process, all the participants strongly agreed that their stories were represented as they intended.

A litmus test of the validity of the method utilised is how participants experience their engagement. In such contexts, I have found that students, more so than staff, have responded favourably to such methods. For instance, in a questionnaire asking about their reception of the visual narrative evaluation method (discussed in 3.2), all the participating students indicated it was preferable to any other method of evaluation they had experienced, because its reflective component incorporated the affect. The majority of those students indicated that their participation in the process increased their intrinsic motivation in their studies (Meistre & Belluigi, 2010).

In terms of the more open, dialogical manner of the research methods employed in 3.2, staff participants appreciated the opportunities to reflect on and debate a problematic central to their own internal conundrums in their professional and teaching practice. A minority (2 of the 14 participating staff) noted fatigue in the intensive nature of the interactions during the process, even though one later emailed that it had been "like a learning curve, demanding but ultimately good". Similarly, student participants were appreciative of having more ownership of the process, requesting I share the various ways in which the research was disseminated, to which many have continued to responded with questions, comments and further additions.

7.4 Conclusion

Both in my own research and teaching practice, and in what I have encouraged in the evaluation practice of my peers, the impetus has been to go beyond "describe the parts... but also to understand relational and contextual factors" of the problematics at hand, a feature characteristic of arts-based educational research practices (Sullivan, 2014, p. 9). The approaches described in this chapter are drawn from and across a diversity of research methods and sources, much like a *bricoleur* (Hickman, 2008). Such eclectic methodological triangulation has been to develop multi-perspectival, dimensional and layered representations of the significance of approaches to teaching, learning and assessment in HE. In this process, warranted assertability of the research process was ensured.

Moreover, in this chapter I have the concerted efforts I continue to make in an endeavour to create conditions for transgression of traditional positioning in research processes, with deliberate opportunities for participants to utilise their agency to not only author their own stories, but have considerable power over how these stories are negotiated during the processes of formation and interpretation, and increasingly the ways in which they were ultimately represented when disseminated. It is this concern with subjectivity and enabling agency within representations of experience, particularly of those oppressed, which I believe is more important than the ongoing debate about the legitimacy of arts-based methods, and it is this ethical obligation which spurs me on to consider ever more diverse approaches.

References

Adorno, T. W. (2004). *Aesthetic Theory*. London: Continuum.

Alcock, A. (2017). *Positioning 'the self': Comparative Case Studies of First Generation Students' Academic Identities When Home Meets Campus in a Rapidly Transforming Higher Education Context*. Grahamstown: Rhodes University.

Bain, W. (1995). "The loss of innocence: Lyotard, Foucault, and the challenge of postmodern education," in *Education and the Postmodern Condition,* ed. M. Peters. London: Bergin & Garvey.

Barnett, R. (2000). University knowledge in an age of supercomplexity. *High. Educ. 40*, 409–422. doi: 10.1023/A:1004159513741

Barone, T., and Eisner, E. W. (1997). "Arts-based educational research," in *Complementary Methods for Research in Education* 2nd Edn, ed. R. M. Jaeger. (Washington, WA: AERA), 73–116.

Barthes, R. (1984). *Camera Lucida: Reflections on Photography*. London: Fontana.

Belluigi, D. Z. (2001). *Broken Vessels: The Im-Possibility of the Art of Remembrance and Re-Collection in the Work of Anselm Kiefer, Christian Boltanski, William Kentridge and Santu Mofokeng*. Masters thesis, Rhodes University, Grahamstown.

Belluigi, D. Z. (2008a). *Excavating the 'Critique': An Investigation into Disjunctions between the Espoused and the Practiced within a Fine Art Studio Practice curriculum.* Masters thesis, Rhodes University, Grahamstown.

Belluigi, D. Z. (2008b). "Making allowance for doubt: arguing for validity in evaluation processes," in *Understanding Learning-Centred Higher Education*, eds C. Nygaard, and C. Holtham (Copenhagen: Calibri Press).

Belluigi, D. Z. (2009). Exploring the discourses around 'creativity' and 'critical thinking' in a South African creative arts curriculum. *Stud. High. Educ.* 34, 699.

Belluigi, D. Z. (2011). Intentionality in a creative art curriculum. *J. Aesthet. Educ.* 45, 18–36.

Belluigi, D. Z. (2013a). A Proposed schema for the conditions of creativity in fine art studio practice. *Int. J. Educ. Arts*, 14, 1–22.

Belluigi, D. Z. (2013b). "Playing broken telephone with student feedback: the possibilities and issues in transformation within a South African case of a collegial rationality model of evaluation," in *Enhancing Learning and Teaching through Student Feedback in Social Sciences*, eds S. Nair and P. Mertova (Cambridge: Woodhead Publishing).

Belluigi, D. Z. (2016). *Evaluation of Teaching and Courses: Reframing Traditional Understandings and Practices*. Grahamstown: Rhodes University.

Belluigi, D. Z. (2017a). A framework to map approaches to interpretation. *J. Aesthet. Educ.* 51, 91–110.

Belluigi, D. Z. (2017b). The importance of critical judgment in uncertain disciplines: a comparative case study of undergraduate fine art visual practice. *Arts Human. High. Educ.* [Published online on June 4, 2017].

Belluigi, D. Z., and Thondhlana, G. (2016/2017). *Research Project: Considering the Influence of Individual and Group-Based Identities on Recipients' Reception of Equity-Agenda Development Programmes in a Rapidly Transforming Higher Education Context.*

Bengtsen, S., and Barnett, R. (2017). Confronting the dark side of higher education. *J. Philos. Educ.* 51, 114–131.

Besnier, N. (1990). Language and affect. *Annu. Rev. Anthropol.* 19, 419–451.

Bessell, A. G., Deese, W. B., and Medina, A. L. (2007). Photolanguage how a picture can inspire a thousand words. *Am. J. Eval.* 28, 558–569.

Bleakley, A. (1999). From reflective practice to holistic reflexivity. *Stud. High. Educ.* 24, 315.

Boulton, A., Grauer, K., and Irwin, R. L. (2017). Becoming teacher: a/r/tographical inquiry and visualising metaphor. *Int. J. Art Des. Educ.* 36, 200–214.

Brookfield, S. (1995). *Becoming a Critically Reflective Teacher*. San Francisco, CA: Jossey-Bass.

Buck-Morss, S. (1997). *The Dialectics of Seeing: Walter Benjamin and the Arcades Project*. Cambridge, MA: MIT Press.

Burgin, V. (1982). *Thinking Photography*. Basingstoke: Palgrave Macmillan.

Cabrera, M., and Guarln, O. (2012). Image and the social sciences. *Mem. Soc.* 16, 7–15.

Chappell, C., Rhodes, C., Solomon, N., Tennant, M., and Yates, L. (2003). *Reconstructing the Lifelong Learner: Pedagogy and Identity in Individual, Organisational and Social Change*. London: Routledge.

Clarke, G. (1997). *The Photograph*. Oxford: Oxford University Press.

Cook-Sather, A., and Alter, Z. (2011). What is and what can be: how a liminal position can change learning and teaching in higher education. *Anthropol. Educ. Q.* 42, 37–53.

Cronin, O. (1989). "Psychology and photographic theory," in *Image-Based Research. A Sourcebook for Qualitative Researchers*, ed. J. Prosser (London: RoutledgeFalmer).

Dallow, P. (2003). Representing creativeness: practice- based approaches to research in creative arts. *Art Des. Commun. High. Educ.* 2, 49–66.

Davies, B., and Hare, R. (1990). Positioning: the discursive construction of selves. *J. Theory Soc. Behav.* 20, 43–63.

Day Sclater, S. (2003). What is the subject? *Narrative Inq.* 13, 317–330.

Derrida, J. (1981). *Writing and Difference*. London: Routledge & Kegan Paul.

Diamond, C. T. P., and Mullen, C. A. (1999). *The Post-Modern Educator*. New York, NY: Peter Long.

Dirkx, J. (2001). The power of feelings: emotion, imagination, and the construction of meaning. *New Dir. Adult Contin. Educ.* 89, 63–72.

Edwards, D. (2007). Restructuring implicational meaning through memory-based imagery: Some historical notes. *J. Behav. Ther.* 38, 306–316.

Edwards, D. (2011). "Invited essay: from ancient shamanic healing to 21st century psychotherapy: the central role of imagery methods in effecting psychological change," In *Oxford Guide to Imagery in Cognitive Therapy*," eds A. Hackmann, J. Bennett-Levy, and E. Holmes (Oxford: Oxford University Press).

Foster, H. (1985). *Postmodern Culture*. London: Pluto Press.

Foucault, M. (1979). *Discipline and Punish: The Birth of the Prison*. (Harmondsworth: Penguin).

Foucault, M. (1980). "Two lectures," in *Power/Knoweldge*, ed. C. Gordon (New York, NY: Pantheon Books).

Foucault, M. (1984). "What is enlightenment?," in *The Foucault Reader*, ed. P. Rabinow (New York, NY: Pantheon Books).

Friere, P. (1972). *The Pedagogy of the Oppressed*. Harmondsworth: Penguin.

Gebauer, G., and Wulf, C. (1995). *Mimesis: Culture, Art, Society*. Berkeley, CA: University of California Press.

Giroux, H. A. (1988). *Schooling and the Struggle for Public Life: Critical Pedagogy in the Modern Age*. Minneapolis, MN: University of Minnesota Press.

Grundy, S. (1987). *Curriculum: Product or Praxis?* London: Falmer Press.

Hickman, R. (2008). "The nature of research in arts education,". In *Research in Art and Design Education,* ed. R. Hickman (Bristol: Intellect Books).

Holloway, W., and Jefferson, T. (2000). "Narrative, discourse and the unconscious: The case of Tommy," in *Lines of Narrative: Psychosocial Perspectives,* eds S. Day Sclater, C. Squire, and A. Treacher (London: Routledge).

Jansen, J. (2008). *Bitter Knowledge – University World News*. Available at: http://www.universityworldnews.com/article.php?story=20080320161054152 [accessed March 3, 2017].

Kalin, N. M., and Barney, D. T. (2014). Hunting for monsters: visual arts curriculum as agonistic inquiry. *Int. J. Art Des. Educ.* 33, 19–31.

Kaplan, R. M., and Saccuzzo, D. P. (2012). *Psychological Testing: Principles, Applications, and Issues,* 8th Edn. Belmont, CA: Wadsworth Publishing Co Inc.

Kristeva, J. (1991). *Strangers to Ourselves*. Hemel Hempstead: Harvester Wheatsheaf.

Lather, P. (1993). Fertile Obsession: Validity after Poststructuralism. *Sociol. Q*. 34, 673–693.

Lather, P. A. (1991). *Feminist Research in Education: Within/Against*. Waurn Ponds, VIC: Deakin University.

Leonardo, Z. (2004). The color of supremacy: beyond the discourse of 'white privilege'. *Educ. Philos. Theory* 36, 137–152.

MacLure, M. (2003). *Discourse in Educational and Social Research*. Philadelphia, PA: Open University.

Mann, S. (2001). Alternative perspectives on the student experience: alienation and engagement. *Stud. High. Educ.* 26, 8–19.

Meistre, B. (1998, ongoing). *Malaise. Growing Photographic Archive of 10 x 15 cm Photographic Prints*. Collection of the artist.

Meistre, B., Belluigi, D. Z. (2010). "After image: using metaphoric storytelling," in *Teaching Creativity – Creativity in Teaching*, eds C. Nygaard, N. Courtney, and C. Holtham (Oxfordshire: Libri Press).

Meistre, B., and Knoetze, J. (2005). Maliaise: a projective (non-test). *Presented at the International Society for Theoretical Psychology Conference*, UCT Business School, Cape Town.

Miles, M. (2006). "Postmodernism and the art curriculum: a new subjectivity," in *Art Education in a Postmodern World: Collected Essays*, ed. T. Hardy (Bristol: Intellect Books).

Nygaard, C., and Belluigi, D. Z. (2011). A proposed methodology for contextualised evaluation in higher education. *Assess. Eval. High. Educ.* 36, 657–671.

Parker, J. (2003). Reconceptualising the curriculum: from commodification to transformation. *Teach. High. Educ.* 8, 529–543.

Parker, S. (1997). *Reflective Teaching in the Postmodern World: A Manifestofor Education in Postmodernity*. Buckingham: Open University Press.

Peters, M. (ed.). (1995). *Education and the Postmodern Condition*. New York, NY: Bergin & Garvey.

Prosser, J., and Prosser, J. (1998). "The status of image-based research," in *Image-Based Research*, ed. J. Prosser (London: RoutledgeFalmer).

Prosser, J., and Schwartz, D. (1998). "Photographs within the sociological research process," in *Image-Based Research*, ed. J. Prosser (London: RoutledgeFalmer).

Ptaszynski, M., Rzepka, R., Araki, K., and Momouchi, Y. (2011). "Research on emoticons: review of the field and proposal of research framework," in *Proceedings of the Conference: The Seventeenth Annual Meeting of The Association for Natural Language Processing*, Toyohashi, Vol. NLP-2011, 115–116.

Rifà-Valls, M. (2011). Experimenting with visual storytelling in students' portfolios: narratives of visual pedagogy for pre-service teacher education. *Int. J. Art Des. Educ.* 30, 293–306.

Roberts, P. (2007). Intellectuals, tertiary education and questions of difference. *Educ. Philos. Theory* 39, 480–493.

Said, E. (1993). "The politics of knowledge," in *Race, Identity, and Representation in Education*, ed. C. McCarthy and W. Crichlow (New York, NY: Routledge).

Seddon, D. (2014). Be a mighty hard message': toni morrison's beloved and the exploration of whiteness in the post-apartheid classroom. *Safundi* 15, 29–52.

Shorr, J. E. (1983). *Psychotherapy Through Imagery,* 2nd Edn. New York, NY: Thieme-Stratton.

Silverman, L. (2007). There are five sides to every story: Which are you missing? *Commun. World* 24. Available at: //www.questia.com/magazine/1G1-156483171/there-are-five-sides-to-every-story-which-are-you

Smith, J. K. (2004). "Learning to live with relativism," in *Educational Research: Difference and Diversity*, I. Stronach and H. Piper (Burlington, VT: Ashgate), 45–58.

Starr-Glass, D. (2005). Metaphors and maps in evaluation. *Assess. Eval. High. Educ.* 30, 195–207.

Stewart, R. (2008). "Constructing neonarratives: a pluralistic approach to research," in *Research in Art and Design Education,* ed. R. Hickman, (Bristol: Intellect Books), 157–163.

Stronach, I., and MacLure, M. (1997). *Educational Research Undone: The Postmodern Embrace*. Philadelphia, PA: Open University Press.

Sullivan, G. (2007). "Creativity as research practice in the visual arts," in *International Handbook of Research in Arts Education*, ed. L. Bresler (Berlin: Springer).

Sullivan, G. (2014). What does it mean to have an N of 1? Art making, education, research, and the public good. *Int. J. Educ. Arts*, 15.

Tagg, J. (1988). *The Burden of Representation: Essays on Photographies and Histories*. Minneapolis, MN: University of Minnesota Press.

Tight, M. (2004). Research into higher education: an a-theoretical community of practice? *High. Educ. Res. Dev.* 23, 395–411.
Van der Poel, N. (2016). "Enabling affective responses within questionnaires," in *Evaluation of Teaching and Courses: Reframing Traditional Understandings and Practices*, ed. D. Z. Belluigi (Grahamstown: The Centre for Higher Education Research, Teaching and Learning), 20–22.
Wallin, J. (2008). Living with monsters: an inquiry parable. *Teach. Educ.* 19, 311–323.
Winnicott, D. W. (1971). *Playing and Reality*. London: Tavistock.

8

Blind Running: 25 Pictures Per Page

Alison Laurie Neilson[1], Andrea Inocêncio[2], Rita São Marcos[1], Rodrigo Lacerda[3], Maria Simões[4], Simone Longo de Andrade[5], Rigel Lazo Cantú[6], Nayla Naoufal[7], Maja Maksimovic[8] and Margarida Augusto[9]

[1]Centre for Social Studies, CES, University of Coimbra, Colégio de S. Jerónimo, Apartado 3087, 3000-995 Coimbra, Portugal
[2]Colégio das Artes, University of Coimbra, Apartado 3066, 3001-401 Coimbra, Portugal
[3]Center for Research Network in Anthropology, CRIA, School of Social Sciences and Humanities, New University of Lisbon/ISCTE University Institute of Lisbon, Av. Forças Armadas, Edifício ISCTE-IUL, 1649-026 Lisboa, Portugal
[4]Descalças Cultural Cooperative, Estação de Caminhos de Ferro de Castelo de Vide, 7320-441 Castelo de Vide, Portugal
[5]Human Rights Consultant, Lisbon, Portugal
[6]Faculty of Social Work and Human Development, Autonomous University of Nuevo León, San Nicolás de los Garza, Nuevo León, México C. P. 66450, México
[7]Department of Teacher Education and School Research, University of Oslo, P.O. Box 1099, Blindern 0317, Oslo, Norway
[8]Department for Pedagogy and Andragogy, University of Belgrade, Studentski trg 1,11000 Belgrade, Serbia
[9]Department of Sociology, Faculty of Economics, University of Coimbra, Avenida Dias da Silva 165, 3004-512 Coimbra, Portugal

Abstract

This experiential visual open work is built from a myriad of words, languages, cultures, and critical theories.

> *. . . books and bombs, dance and record labels, mothers and daughters, small villages and islands, diaries and story, colonization and immigration, violence and healing, leaving and returning . . . wild animals*

It is a collaborative attempt to use untranslated images and untranslated embodied praxis trusting in one another to look out for the enhanced dangers of running to meet gruelling deadlines and unrelenting competition for survival in the academy while at the same time stubbornly resisting, via a blindfold, the reigning forms of knowing and communication.

In the summer of 2016 a group of 13 people, from a multitude of different life experiences gathered for a 4 day retreat from academia held at a seminary.

Our mission was to explore art as a way of knowing.

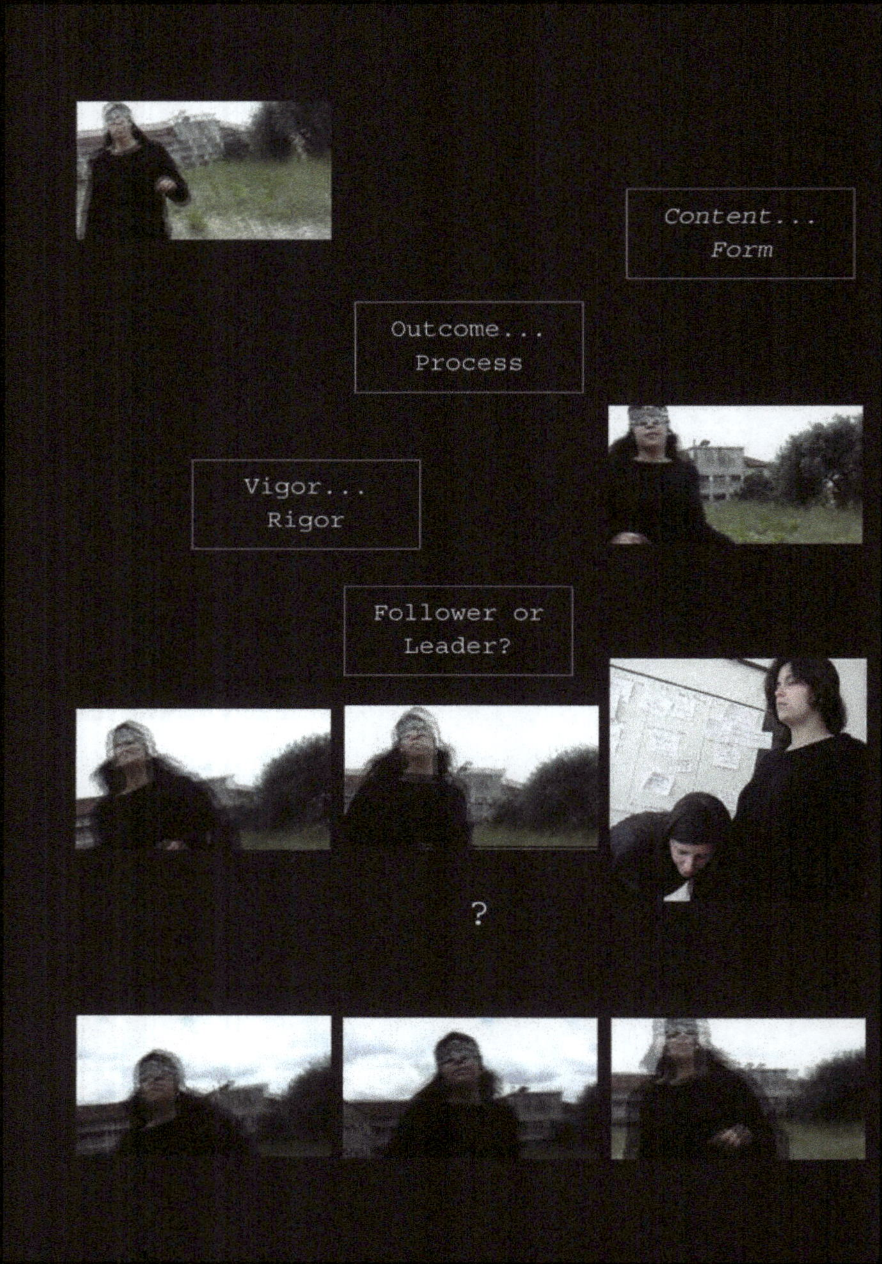
Content... Form
Outcome... Process
Vigor... Rigor
Follower or Leader?
?

WHAT are we doing?

"Run?"
"Run!"
"WATCH OUT!"
"Opps...
not so far"

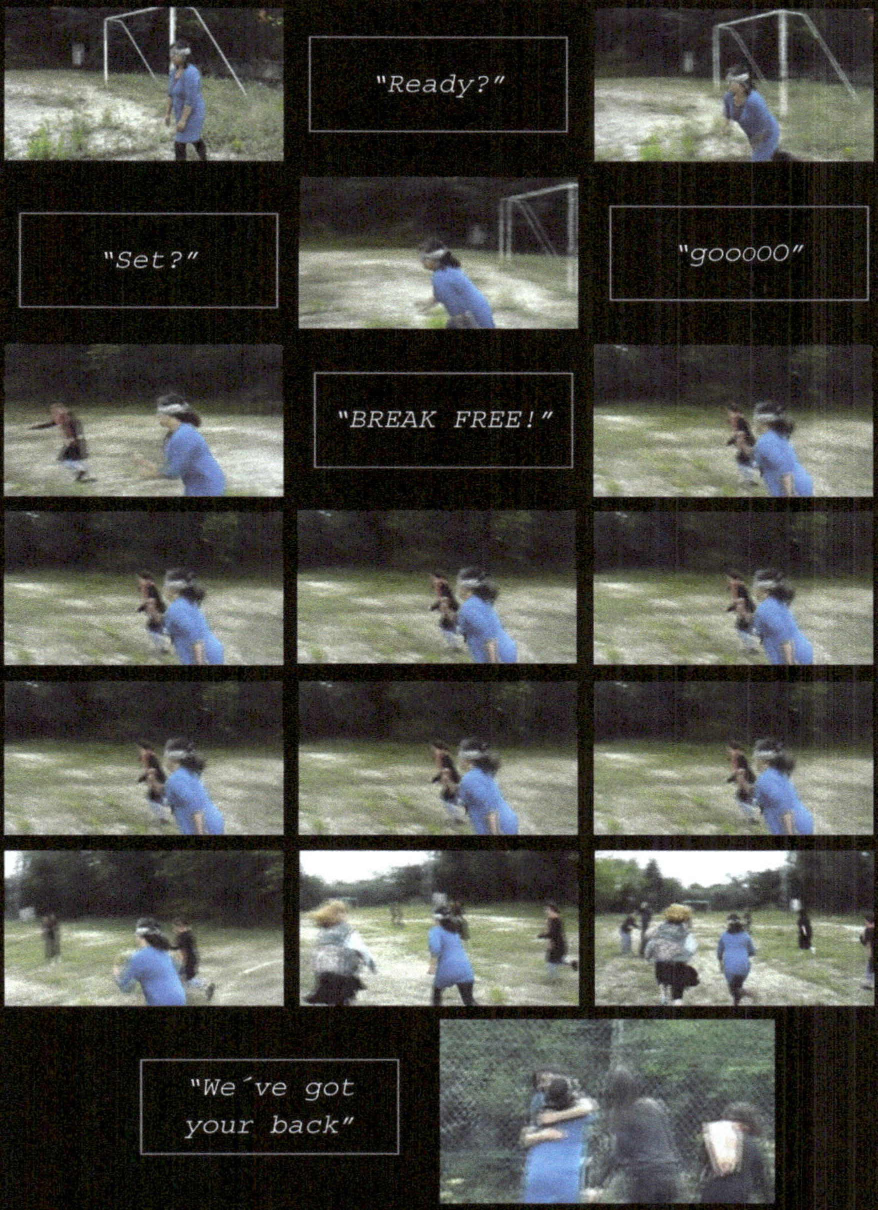
"Ready?"
"Set?"
"gooooO"
"BREAK FREE!"
"We´ve got your back"

Thinking

Feeling

Let´s put ourselves
on the line

Transform and live moments of equity with space for all our sensory perceptions, reasons, logics, intuitions, inspirations and wisdom.

Trust,
just run

"

" (Sousanis, 2015, p. 11).

Dear reader,

If you have been obedient to the norms of left to right, top to bottom reading, you have already made your way through a photo essay, which ended with one of the blindfolded runners escaping the confines of the filmstrip to enter the images of Nick Sousanis (2015) which we have quoted. Did you linger on the images? Did you smile in recognition or resonance; or, did you flip casually through the pages, perhaps with some annoyance at the ambiguity of the presentation, or maybe with relief to get here quickly as your tall pile of other work beckons?

"

" (Sousanis, 2015, p. 62).

"

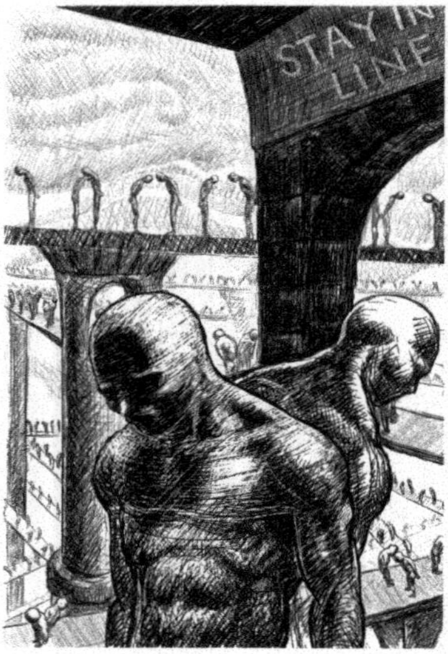

" (Sousanis, 2015, p. 3).

The photo-essay presentation, "Blind running: 25 pictures per page" attempts to meet three objectives, whose value could be contested and which might be in conflict with each other. The first and most normal objective is to communicate about the idea of *blind running* which emerged from a 4-day summer school about ways of knowing. The second is a continuous reflective engagement – a blind running – with all those who have helped create this chapter, and, third is a hope that any ambiguity, strangeness or confusion provoked by the graphic presentations will invite readers to join in blind running as a way to reflect on what your expectations and practices privilege.

Although our original intention was to offer the photo essay without further explicative text, the editors of this book, asked us to touch on our rationale and challenged us to sharpen our communications. To this end, we draw attention to how any of us might identify with the images represented throughout the pages and what might keep us from seeing the lines within which we may or may not want to stay. As this work deals directly with practices of scholarship – researching, publishing, teaching, learning and routinely engaging with editors regarding the specifics of how to communicate in publications such as this chapter, we invite Tatiana and Xyangyun, to run blindly with us as well. With the help of the publisher, we are trying to take up opportunities

"

" (Sousanis, 2015, p. 5).

created by the medium we are using, knowing that fonts and margins can change our presentation, but also aware, like Marshall McLuhan (McLuhan & Fiore, 1967) that the message may still be massaged in ways unexpected by us: a benefit of blind running (cf. Open work, Eco, 1989).

What we are doing with the graphic essay is inviting readers to "try on" and experiment with the idea that their truths, interpretations and reactions to the graphic essay may be influenced by visual myths about teaching, teachers, mind and body (Barthes, 1977, 1981 cited in Hallewell & Lackovic, 2017). Additionally, in asking the reader to do this, we are deliberately taking a blind position. We cannot foresee who reads this, how and what are taken as truth and experimental myths. We give up control, as we have also asked the editors of the book to do.

We are not blind, but yet we are and through the act of deliberately covering our eyes to words, can we not learn to see that which the words most powerfully hide?

"

" (Sousanis, 2015, p. 144).

Chemi (2017) recognized artistic experiences as a place where we can find "a safe haven for cognitive and emotional challenges, for experimentations, for learning and developing, for including heuristics in knowledge, for indirect cognition and communication (metaphors), for training resilience and opportunity-seeking strategies". However, our safety is difficult to protect and we can only run blind without harming or being harmed in higher education, when our colleagues "have our backs", through a mechanism of community of practice (Lave & Wenger, 1991) which is strengthened via the "generative capacities of art, imagination, intuition or playfulness" (Fegan, 2017, p. 134).

Communicating with Images

In the introduction to the handbook of Visual Research Methods, Pauwels (2011) points to a growing number of scholarly journals dedicated to visual images but little integration of the findings and practices between the social sciences, humanities and behavioural sciences. Less than 25% of the handbook's 754 pages have graphic images with the majority concentrated in a handful of chapters. None of the chapters uses images to convey the prime communication (e.g., graphic novels, comics, photo essay).

Photos as illustration are common in much writing; photos as analysis, exploring what the contents of a photo can tell us about an experience is less common, but not unknown yet, a third use of photos, to make an argument, is quite rare (Newbury, 2011). Very few images are published as text within non-art journals (Pinola-Gaudiello & Roldán, 2015). In our search, we have found the images, which we have quoted from a recently published graphic novel/PhD dissertation (Sousanis, 2015) about accessing multiple modes of understanding. We also became aware of the collaboration of Stephanie Jones and James F. Woglom who have been creating comics based education research publications. We contacted these researchers to know how they thought their graphical work should be quoted and found that we would be running blind in this attempt as well.

Therefore, we looked further but found very little research on how photos are used in higher education. Hallewell and Lackovic (2017) researching lecturing practices, pointed toward Posser and Roth (2003) who used a taxonomy of four functions for using photos: decorative, illustrative, explanatory and complementary and who revealed that "the semiotic potential of photographs in lectures is *underused*" (p. 13), suggesting that higher education practitioners could benefit from training in this regard. In the study of psychology lectures, little critical questioning of the photos used was found: a disturbing finding since meaning-making is based partially on "the socio-cultural conditioning surrounding the photograph: who it is viewed by, presented by, created by, where and why is it produced, who are the agents who have intentions and power (Hodge & Kress, 1988)?" (Hallewell & Lackovic, 2017). Our insistence on using blurred photos is to invite an exploration of the semiotic potential of these repetitive film frames, and provoke further critical reflection via our engagement with images in conversation.

Figure 1. Image from Jones and Woglom (2013, p. 173) "I've faced a similar problem of considering quoting from graphica pieces and not wanting to just quote the "words" from the piece, but include the images themselves . . . I think if we just start doing this in our work it will become acceptable practice. Quoting the words seems at best insufficient and at worst, a misrepresentation of the source we quote. Good luck! I would love to see what you create" (personal communication Stephanie Jones, April 10, 2017).

In this short chapter, we hope that we have offered support for those who have taken analogous risks, or who are considering taking them; all of which potentially makes the running field less dangerous and therefore more inclusive. We seek to be agents of our own power in trying to decolonize higher education (Neilson & São Marcos, 2017) using multiple ways of knowing, in particular artful ways, and blind running through the largely uncharted world of graphic conversations (cf. Valdez Ruvalcaba, 2011). In taking this step, we expose our vulnerabilities rather than taking time to hide them and in doing so we trust that others will have our backs.

Acknowledgements

Thank you to all the participants of the CES Summer School *ArtFULL 2016: Relationships of knowing, doing, and being*; the community of the Instituto Missionário Sagrado Coração, particularly Tiago Pereira, Eduardo Pereira and Humberto Martins; plus António de Campos and Anderson Paiva.

References

Chemi, T. (2017). “A safe haven for emotional experiences: perspectives on the participation in the arts,” in *Innovative Pedagogy: A Recognition of Emotions and Creativity in Education,* eds T. Chemi, S. Grams Davy, and B. Lund (Rotterdam: Sense Publishers), 9–25.

Eco, E. (1989/1962). *The Open Work*, trans. A. Cancogni. Cambridge, MA: Harvard University Press.

Fegan, S. (2017). “Imagining the world: creating an artistic community of practice in an academic environment,” in *Implementing Communities of Practice in Higher Education,* eds J. McDonald and A. Cater-Steel (Singapore: Springer), 131–156.

Hallewell, M. J., and Lackovic, N. (2017). Do pictures “tell” a thousand words in lectures? How lecturers vocalise photographs in their presentations vocalise photographs in their presentations. *High. Educ. Res. Dev.* 36, 1166–1180.

Jones, S., and Woglom, J. F. (2013). Graphica: comics arts-based educational research. *Harv. Educ. Rev.* 83, 168–189.

McLuhan, M., Fiore, Q., and Agel, J. (1967). *The Medium is the Massage: An Inventory of Effects*. New York, NY: Random House.

Neilson, A. L. (2008). *Disrupting Privilege, Identity, and Meaning: A Reflexive Dance of Environmental Education.* Rotterdam: Sense Publishers.

Neilson, A. L., and São Marcos, R. (2017). In response to a call. Evoking a keynote. Special issue on neoliberalism in higher education. *Crit. Stud. Crit. Methodol.* 17, 1–3.

Newbury, D. (2011). "Making arguments with images: visual scholarship and academic publishing," in *The Sage Handbook of Visual Research Methods,* eds E. Margolis and L. Pauwels (London: Sage), 651–664.

Pauwels, L. (2011). "Introduction," in *The Sage Handbook of Visual Research Methods,* eds E. Margolis and L. Pauwels (London: Sage), 3–23.

Pinola-Gaudiello, S., and Roldán, J. (2015). "Images in educational research reports: a literature review of educational researcher," in *Proceedings of the III Conference on Arts-based and Artistic Research,* Barcelona, 1–15.

Sousanis, N. (2015). *Unflattening.* Cambridge, MA: Harvard University Press.

Valdez Rubalcaba, S. G. (2011). "Painting by listening: participatory community mural production," in *Community Arts and Popular Education in the Americas,* ed. D. Barndt (Albany, NY: SUNY Press), 102–112.

9

How Can Inspiration Be Encouraged in Art Learning?

Chiaki Ishiguro[1] and Takeshi Okada[2]

[1]Tamagawa University Brain Science Institute, 6-1-1 Tamagawagakuen, Machida, Tokyo 194-8610, Japan
[2]The University of Tokyo, Graduate School of Education, 7-3-1 Hongo, Bunkyo-ku, Tokyo 113-0033, Japan

Abstract

Inspiration has been regarded as an important phenomenon in research into artistic creation and art learning (Tyler & Likova, 2012; Chemi, Jensen & Hersted, 2015). Active art appreciation inspires people and facilitates the creativity of art-making, as shown by psychological experiments on drawing (Okada & Ishibashi, 2016). Our recent studies on inspiration in artistic activities have shown that the core of inspiration through art appreciation (ITA) is a dual focus on the artwork (and artist) and the viewer's own art-making. In this chapter, we outline our model of the psychological process of ITA. We list factors that promote ITA, in particular, those inducing a dual focus in the context of educational practice. Some factors – instruction, methods of appreciation, and methods of selecting and showing artworks – may contribute to educational interventions in museums and schools. Finally, we described a case of art educational practice for undergraduates, designed to promote inspiration.

9.1 Introduction

Many artists have described receiving inspiration for their creations. In the book, "Behind the Scenes of Artists' Creations", Tatiana Chemi, Julie Borup Jensen and Lone Hersted provided several examples of artists' creative inspiration (Chemi et al., 2015). For example, Julia Varley, an Italian actress, was deeply inspired by a performance of Odin that she had the opportunity to see when visiting Denmark. Although she encountered many difficulties performing in a foreign country, she was so inspired by Odin's work that it pushed her to become a better artist herself. Such an experience has often been described as a gift from the creative muse, as if inspiration comes from some higher power. In fact, psychological research in past decades has revealed that the creation of artwork requires conscious effort (Weisberg, 2006). More recently, researchers have begun to examine inspiration empirically and have suggested that inspiration plays an important role in creativity "by firing the soul" (for a review, see Oleynick, Thrash, LeFew, Moldovan & Kieffaber, 2014). The increase in the number of psychological studies on inspiration in creation has helped us begin to unravel the mystery of inspiration. In addition, researchers studying art learning have also shown an interest in inspiration and how the inspiration process can be utilized in art learning, a question that still remains unanswered (Tyler & Likova, 2012).

To answer this question, this chapter provides a review of psychological studies on inspiration, especially in artistic activities, and suggests important factors in promoting inspiration in art learning and art education. In addition, we introduce our own practice in art education and report how it was designed to promote inspiration and how learners changed throughout the practice. We aim to offer a new framework for art learning, with a focus on inspiration.

9.2 A Brief Review of Psychological Studies on Inspiration

Inspiration has recently become a topic of empirical investigation in psychological studies. The word "inspiration" has been used in various areas such as social comparison, religion, problem solving, and creativity. The triggers, the results, and the processes of inspiration differ depending on the area of study. Two American psychologists studying motivation, Todd M. Thrash and Andrew J. Elliot, provided a general conceptualization of inspiration by focusing on the psychological experience (Thrash & Elliot, 2003, 2004). Through questionnaire survey studies, they statistically extracted the following three elements as the psychological constructs of inspiration: evocation (e.g., feeling overtaken, uncontrolled); motivation (e.g., activation, energy); and transcendence (e.g., positivity, enhancement and clarity). Further, they defined the process of inspiration as being inspired by, which refers to appreciation of the perceived intrinsic value of a stimulus object, or inspired to, which refers to motivation to actualize or extend the valued qualities to a new object. Indeed, they enabled the measurement of the psychological experience of inspiration by developing a psychological scale of inspiration called the "Inspiration Scale (Thrash & Elliot, 2003)".

Thrash and his colleagues also conducted empirical research into inspiration during creative activities. In one study, undergraduates participated in various writing tasks, such as scientific writing and fictional writing, and reported their level of inspiration during the writing process. The results showed that how inspired they felt predicted how creative their works were rated to be by readers. In addition, they often felt inspired *after* they came up with a new idea. These findings contributed to the understanding of the psychological mechanism of inspiration in creation.

However, inspiration might not result from an inner process. As described above, Thrash & Elliot believed that inspiration includes the process of being inspired by *something* (Thrash & Elliot, 2004). Recent studies have pointed out that artists are strongly influenced by encounters with the outside world, and they often actively make use of these encounters when creating art (Chemi et al., 2015; Takagi, Kawase, Yokochi & Okada, 2015). We gain stimulation from outside phenomena, such as an apple falling to the ground, traces of our own scribbles, artworks created by amateurs or famous artists, and incidents in our personal or social lives. Especially in art education and art learning, it is important that students learn from artworks created by others and gain inspiration from them. According to the Systems Model of Creativity (Csikszentmihalyi, 1999), learners can participate in the social system of creativity by acquiring knowledge and inspiration from other creators and their works.

Focusing on such theories and episodes, some researchers have claimed that inspiration occurs as a result of encounters with the world outside the self. For example, analogy researchers have examined the relationship between one's creativity and one's encounters with the outside world. They define analogy as transferring meaning or ideas from a particular subject (analogical source) to another (target), and express the similarity or difference based on the distance of the source and target. For the explanation of creativity, they focused on encounters with images, products, and ideas as the analogical source, and proposed that a conceptually distant source, which has a similar structure but a different surface, elicits a creative breakthrough (Conceptual leap hypothesis: Gentner & Markman, 1997; Holyoak Thagard, 1996; Poze, 1983; Ward, 1998). The hypothesis was confirmed in studies on product design in artistic domains (Chan, Schunn, Cagan, Wood, Kotovsky, 2011). However, results of more recent studies implied that the conceptual distance of sources did not affect the novelty of ideas analogized from the sources (Fu, Chan, Cagan, Kotovsky & Schunn, 2011). In keeping with these studies, Chan et al. suggested that creativity of ideas might be affected by a deep exploration of distant sources (Chan, Dow & Schunn, 2015).

Deep exploration was also examined in an experimental study on drawing (Okada & Ishibashi, 2016). Okada & Ishibashi conducted experiments to study the creative process of drawing. On the first day of Experiment 1, non-art major undergraduates were asked to create a drawing of natural motifs (e.g., pine and pepper). On the second day (the intervention session), one group of participants copied an artist's drawing with a similar motif, while a control group of participants drew motifs in the same manner as on the first day. All participants were asked to create a drawing of natural motifs again on the third day. In Experiment 2, the participants were presented with two types of drawings by artists (e.g., representative and abstract), which were selected according to how familiar they were to the participants (i.e., representative drawing familiar and abstract unfamiliar). In Experiment 3, one group of participants copied an artist's drawing while another group only viewed the artist's drawing for approximately 20 minutes on the second day. The results of these experiments showed that both copying an unfamiliar artwork and viewing it for a long time promoted creativity, while copying familiar artwork did not have that effect. In addition, the participants were also asked to express aloud what they were thinking during the intervention session. The results suggested that they relaxed their constraints (i.e., their preconceptions about art-making) and reconstructed their knowledge through deep exploration of the artwork. These findings provide useful insight into the mechanism of the creative process by stimulation from outside. Recently, Okada (2016) further emphasized the importance of comparison between oneself and others in artistic creation. Through the process of comparison, viewers can detect the differences between their own schemas and others' schemas through profound encounters such as copying others' works or spending a long time appreciating the works of others. In order to examine the importance of the comparison process empirically, we conducted another study, which was a questionnaire survey of non-art major undergraduates. It revealed that appreciation with comparison between one's own art-making and works by others promoted artistic inspiration more strongly than appreciation with evaluation of others' artworks (Ishiguro & Okada, 2015).

Another question is how viewers compare others and themselves while appreciating a work of art. To answer this question, we introduce an outline of the psychological model of inspiration for art-making through art appreciation (the ITA model), and in the next section, we suggest factors to promote ITA in art learning. Though this model primarily focuses on inspiration for art-making through art appreciation, it can be applied to other types of social encounters with the world beyond the self.

9.2.1 Psychological Model of Inspiration for Art-Making through Art Appreciation (ITA)

Previous studies on art appreciation have assumed the goal of art appreciation to be evaluating and understanding artworks (Pelowski, Markey, Lauring & Leder, 2016). However, such studies did not explain how people drew inspiration from others' artworks while they were appreciating a work of art. As we have mentioned above, drawing inspiration from others' artworks is one of the key processes of artistic creativity and needs to be incorporated into the model of art appreciation. Our previous studies suggested that the process of inspiration is triggered by individuals when they compare others' creative process with one's own (Okada, 2016; Ishiguro & Okada, 2015). Therefore, the ITA model needs to include the process of viewers' own art making. The model should extend the process to include both evaluation of artworks and reflection on the viewer's own art-making as its essential goals. In addition, the model assumes that the process of ITA consists of both emotional and cognitive processes. This is because inspiration has been conceptualized as a motivational state, which consists of emotion, cognition, and need (Tyler & Likova, 2012; Thrash & Elliot, 2003, 2004). As shown in the outline of the ITA model (Figure 9.1), the process consists of four phases: the initial state; Phase 1; Phase 2; and Phase 3. The initial state refers to the motivational state for artistic activities such as art-making and appreciation. Phase 1 refers to

Subject of processing	Type of processing	Phases of appreciation for inspiration			
		Initial state	Phase 1	Phase 2	Phase 3
Viewer's own art-making	Cognitive Emotional	Motivational state for art-making		Reflection on viewer's own art-making	Inspiration for viewer's own art-making
Others' artworks	Cognitive Emotional	Motivational state for appreciation	Evaluation of others' artworks	Evaluation of others' artworks	

Figure 9.1 Outline of the process model of inspiration to make artworks through art appreciation (ITA).

evaluation of others' artworks. Phase 2 includes a comparison between the process employed by others and that employed by the viewer of the artwork, based on an evaluation of others' artworks and reflection on the viewer's own art-making. Phase 3 refers to the state of being inspired in art-making. The ITA model emphasizes the significance of Phase 2 because it connects the activities of art appreciation and art-making.

Phase 2 is characterized by its "dual focus", the state of focusing on both the evaluation of others' artworks and reflection on the viewer's own art-making, simultaneously or alternately. To achieve a dual focus, there are some essential conditions. First, it is important for the viewer to be motivated to make art in the initial state. Second, the viewer has to consider how relevant other artist's artwork is to his/her own art-making in Phase 1. The higher the relevance is considered to be, the easier it is for the viewer to pay attention to his or her own art-making and to have a dual-focus.

This feature of the ITA model is useful in order to understand the difference in the experience of inspiration between professional artists and amateurs. It has been shown that experts in a creative art domain (such as design) gain inspiration more frequently and intensely than amateurs (Thrash & Elliot, 2003). Artists are more easily inspired to make art because they are often motivated to make art in the initial state before they enter the stage of appreciating others' works. In addition, they tend to consider how relevant the works by others to their own art-making because they are strongly motivated to improve their creativity. In contrast, amateurs in artistic activities hardly gain any inspiration because when viewing others' artworks they are motivated to appreciate rather than to make art. As a result, their appreciation is often preoccupied with evaluating others' works. However, as shown in an experiment by Okada & Ishibashi (2016), their creativity can be promoted if they actively interact with works by others for a long period of time. We can assume that such deep interactions with artworks encourage viewers to consider how relevant the works to their own art-making.

9.3 Factors That Promote Inspiration for Art-Making through Art Appreciation in Educational Settings

As mentioned above, recent studies have provided findings demonstrating the psychological mechanism of inspiration in artistic creation. If these findings can be applied to educational practices in museums, schools, and lifelong learning settings, learners may be able to experience inspiration more frequently and intensely, become motivated to their artistic activities, and be more committed to art culture. Such an experience plays an extremely important role, not only for the creation of experts in an artistic domain, but also for the well-being of amateurs in their creative lives.

Therefore, we will describe what factors affect the emergence of inspiration by referring to our ITA model. According to this model, it is easy for a viewer of artworks to experience inspiration for art-making if (s)he has a high motivation for art-making in the initial state. However, the model also emphasizes the importance of comparison between the artwork of others and the viewer's own art-making in Phase 2, which includes a dual focus state activating both the attention to an artwork or the artist who made the artwork and the attention to the viewers' own art-making. The dual focus in Phase 2 is also crucial for attaining an inspirational state in Phase 3, whether or not viewers are highly motivated to make artwork. Thus, to promote the process of dual focus, it is essential for novice viewers of art to be creative because, in general, they focus only on an artwork or the artist who made the artwork during art appreciation, and hardly focus on their own art-making.

Therefore, to activate the viewers' focus on their own art-making in an art educational setting, teachers need to instruct viewers on how to interact appropriately with an artwork. A teacher also needs to select appropriate artworks for the students. Figure 9.2 shows examples of interventions to activate a viewer's dual focus: the focus on the artwork and artist, and on the viewer's own art-making. The focus on the artwork and the artist refers to interpreting and evaluating the artwork and the artist. The focus on the viewer's own art-making means reflection on the knowledge, abilities and autobiographical memory of the viewer's own art-making. The examples of the intervention are classified into three ways of interacting with an artwork. The following describes examples of these and how they drive viewers to be dual-focused and inspired.

9.3.1 Interventions

The first type of intervention is to activate viewers' attention to their own art-making process. For this purpose, the most direct intervention is to support viewers in developing their goals for art-making during their art appreciation. For example, it is useful to let viewers first make their own artworks before appreciating others' artwork and then to give explicit instructions such as "appreciate an artwork in order to obtain hints for your own art-making". Researchers have applied this practice of having participants make products

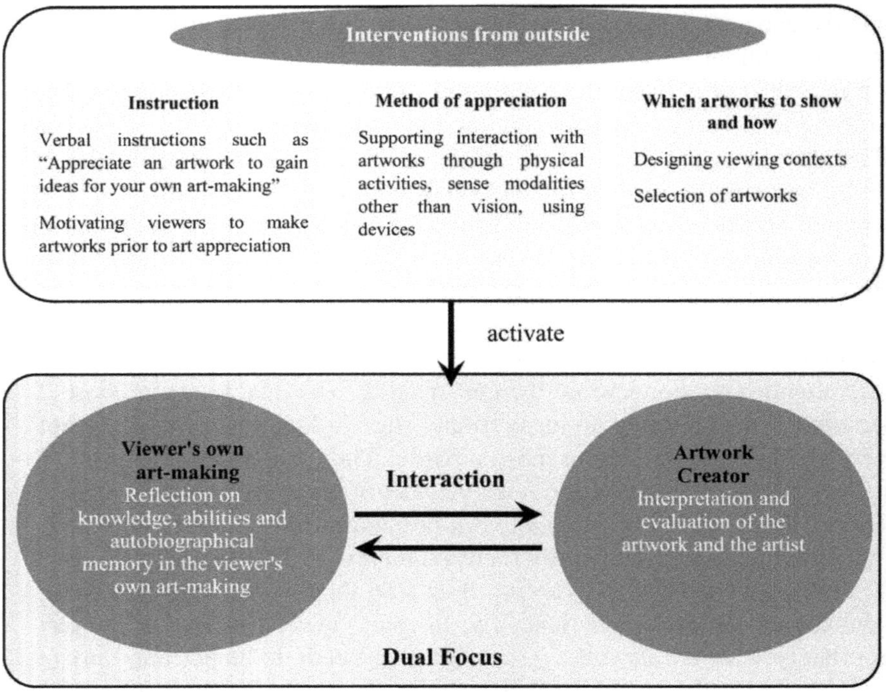

Figure 9.2 Interventions from outside to make viewers dual-focused.

before viewing others' examples in order to investigate creative problem solving, such as product design and artistic drawing (Chan et al., 2011; Langer, Pirson & Delizonna, 2016; Kiyokawa, Izawa & Ueda, 2006). The findings have shown that such interventions promote creative performance and self-evaluation.

The second type of intervention is the method of art appreciation. We describe three methods: appreciation with physical activities; appreciation with sense modalities other than vision; and appreciation using devices. All of these methods of art appreciation are accompanied by actions other than viewing, which is the central activity of art appreciation. These methods make it possible for viewers to experience what they usually do not perceive in everyday life and to question their own preconceptions. If they come to pay attention to themselves during art appreciation, especially to their own art-making process, and they also consider artwork as being relevant to their own art-making, then the comparisons in Phase 2 will be made.

The first method for art appreciation is appreciation with physical activities when viewing artworks. This changes the ways in which viewers' interpret an artwork and helps to direct their attention to themselves. There can be various interventions to promote such appreciation. We present three examples, which have already been applied to learning in practical situations such as in museums and schools. It is suggested that these examples were effective in promoting viewers' dual focus (both interpreting others' artworks and reflecting the viewers' senses, actions and their own feelings). For instance, Nakano & Okada (2016) reported a two-day educational practice in a museum, in which participants appreciated one of the three official replicas of Marcel Duchamp's famous modern art, 'The Bride Stripped Bare by Her Bachelors, Even (often called The Large Glass)', and then they did dance exercises and created their own performance in front of the artwork. They suggested that dancing while appreciating an artwork enabled the participants to fix their attention on both the artwork and their own bodies.

The second method is art appreciation using sense modalities other than vision, such as the sense of touch. This also serves the function of focusing the viewers' attention on their own senses. In general, vision is the dominant sense in aesthetic appreciation of the fine arts. However, art appreciation using other sense modalities provides viewers with new aspects of interpretation and reminds them to pay closer attention to their own senses.

The third method is art appreciation using a device, such as a virtual reality (VR) device. Recently, researchers demonstrated that using VR in museums and galleries enabled viewers to perceive expression in a painting intuitively by interacting with it in a 3D world (Huang & Han, 2014). VR has already been used in art museums; for example, the MoMA in New York has started to utilize VR technology in educational exhibitions. Although there are still few studies about the effect of VR in museums, VR would change our art experience from passively viewing artworks to actively playing with them. Such an experience makes it possible for viewers to go beyond the boundary of their ordinary feelings and thinking and, as a result, to reflect on themselves more actively.

The third type of intervention relates to which artworks to show to viewers and how. Educators and researchers can intervene in how viewers appreciate artworks and what kinds of artworks they consider. The environment and context are thought to be important for art appreciation. Recent studies on art appreciation have suggested that the evaluation and interpretation of an artwork is affected by its environment and context (Leder, Belke, Oeberst & Augustin, 2004; Bullot & Reber, 2013) because the context activates viewers' specific knowledge and memories and guides their interpretation of artworks. Especially if the activated knowledge and memories are related to their own art-making, such a context should increase their motivation to make art.

In addition, selecting what kind of artworks viewers should appreciate plays an important role. Depending on what kind of artistic characteristics the viewers perceive, their evaluation and interpretation of the artwork during the appreciation phase may differ. For inspiration, it is important whether or not the perceived characteristics of the artwork are related to the viewers' own art-making.

How do viewers then consider how relevant an artwork to their own art-making? They consider the relevance of the artwork to themselves and their own art by applying their own knowledge and experience. This process is thought to be analogical, with related knowledge and experience being mapped onto the artwork. Literature on art appreciation claims that the style and content of artworks are categorized on the basis of viewers' prior knowledge and experience (Leder et al., 2004). Such a process is related to analogical thinking. Therefore, in the next paragraph we will explain the art appreciation process according to theories of analogy.

Accumulated research into analogy reveals that the core of the human thinking process is deeply related to the perception of similarity and difference. One of the main findings is that analogical processing rests on the common structure between source and target (structural mapping theory: (Gentner, 1983), and it is also influenced by pragmatic contexts (Gick & Holyoak, 1980; Holyoak & Koh, 1987). In contrast, everyday thinking, including creative thinking, often has no clear source and target. Gentner & Markman (1997) proposed a theory known as the structural alignment theory, which states that a new structure can be produced by comparing a source and a target. We can assume that the same process occurs in art appreciation, in the early phase in which viewers classify the content and style of an artwork by applying their prior knowledge and experience implicitly and explicitly (Leder et al., 2004). According to the structural alignment theory (Gentner & Markman, 1997), art appreciation can be regarded as a process of interpreting an artwork (target) on the basis of prior knowledge and experience (source), in which the comparison between the target and source will produce a new structure for alignment. If viewers' autobiographical knowledge of art-making is involved as a source in the process of structural alignment, they consider the artwork as relevant to their own art-making. As a result, the viewers will become dual-focused.

According to these findings, educators should take into consideration learners' prior knowledge and experience of art-making when selecting artworks for their appreciation. To facilitate their dual focus process, it would be better to make the relevance between the artworks and learners' knowledge and experience high. However, if educators intend to make students not only inspired to make art but also to be creative, it might be better to select artworks unfamiliar artworks to the learners, as Okada & Ishibashi (2016) suggested. What kinds of artworks should be chosen depends on the goal of the educational practice.

9.4 An Example of Educational Practice for Promoting Inspiration

In the last half of this chapter, we introduce our practice as an example of supporting learners to gain inspiration through art education.

We conducted a fine-art photography course, "Artistic Creation", at the University of Tokyo, in Japan. The course was held for the cultivation of students' creative fluency or creative literacy, which refers to an understanding of creative processes and methods and the acquisition of habits and attitudes to enjoy creative activities (Agata & Okada, 2013). Creative fluency is assumed to play an important role in individuals' well-being, especially in creative life, and their participation in a creative society. For this purpose, the course was aimed at promoting art-novices' commitment to artistic activities in a certain domain. Learning through experiencing inspiration by the artworks of others is one of the significant parts of an experience of artistic creation. Thus, the educational interventions were designed to encourage such inspiration.

All of the students who participated in the course for course credit were non-art majors and had never had any formal education in artistic activities. We chose artistic photography as the target domain, because photography is familiar even to such students. It was thought to be relatively easy for them to create artistic works through photography, rather than through other forms of art, such as drawing or painting, because only a limited number of basic techniques must be mastered to take pictures once the students are able to use an automatic single-lens reflex (SLR) camera.

The course consisted of 14 classes in total (see Tables 9.1, 9.2 and Figure 9.3), and 21 undergraduates at the university (10 males and 11 females; aged from 20 to 27, M = 21.33, SD = 1.58) followed the course. All of them were beginners in artistic photography. They used a digital SLR camera provided by us for the course (five of them used their own cameras). The whole

Table 9.1 Schedule of the course

Class 1	Guidance
Class 2	Lecture 1
Class 3	Free photography 1
Class 4	Lecture 2
Class 5	Lecture 3
Class 6	Lecture 4
Class 7	Free photography 2
Class 8	Appreciation & Imitation 1
Class 9	Free photography 3
Class 10	Appreciation & Imitation 2
Class 11	Free photography 4
Class 12	Presentation
Class 13	Free photography 5
Class 14	Introduction of the instructors' artworks

N.B. Pale grey cells refer to the 1st intervention designed for inspiration, and grey cells refer to the 2nd intervention designed for inspiration.

Table 9.2 Educational interventions and timetables

	Free Photography (5 Times)	1st Intervention Designed for Inspiration Lecture (4 times)	2nd Intervention Designed for Inspiration Appreciation & Imitation (twice)	Presentation (Once)
13:00–13:15	Complete a questionnaire Question time with the instructor	Question time with the instructor	Appreciation of an exemplar photograph	Appreciation of photographs by the students
13:15–13:45	Photography	Lecture by the instructor		
13:45–14:20		Photography	Photography (imitation)	Presentation
14:20–14:40	Appreciation of some students' photographs and comments on them by the instructor			
Homework				
Explaining the photographs they had taken in the classes Describing what they had considered and noticed about artistic creation each week				

N.B. The actual timetable was adjusted according to the situation.

course was taught by a professional photographer, Fumimasa Hosokawa, who has an MFA in photography, teaches at a professional school of photography, and has held exhibitions domestically and internationally. The second author

Figure 9.3 Classes.

organized the course with the photographer; the first author collected student data, such as the results of student questionnaires and student homework responses. The first author also conducted interviews with the students one year after the course. The students were informed that the data would be analysed and published as scientific research.

The Course Design

The course was designed on the basis of three factors to promote inspiration, as described in Section 9.3 above (see Figure 9.4). In the course, we included two types of intervention for inspiration, combining each of the following factors: instruction, methods of appreciation, and methods of exhibiting artworks.

First, we held a series of lectures as the initial intervention designed for inspiration, taking into account the fact that the students were beginners in artistic photo expression. With the lectures, the students were able to gain hands-on experience taking photos and acquire knowledge and techniques for artistic photo expression and creation. Specifically, there were 4 lectures: the 1st lecture was an introduction to how to use a digital SLR camera (e.g., basic manipulation of exposure, diaphragm, and shutter speed); the 2nd lecture was on lighting; the 3rd lecture was on approaching models; and the 4th lecture was intended to brush up these skills. Each class included a 20-minute lecture on the topic and a practical photography session using the knowledge and techniques learned in each lecture. The instructor answered students' questions at any time during the lectures, and at the end of the classes, the instructor commented on some of the photographs the students had taken in the practical photography sessions. This intervention was designed to promote the students' motivation for art-making before encountering others' artworks.

The second intervention designed for inspiration, was the 'appreciation and imitation' intervention. We designed the way in which students interact

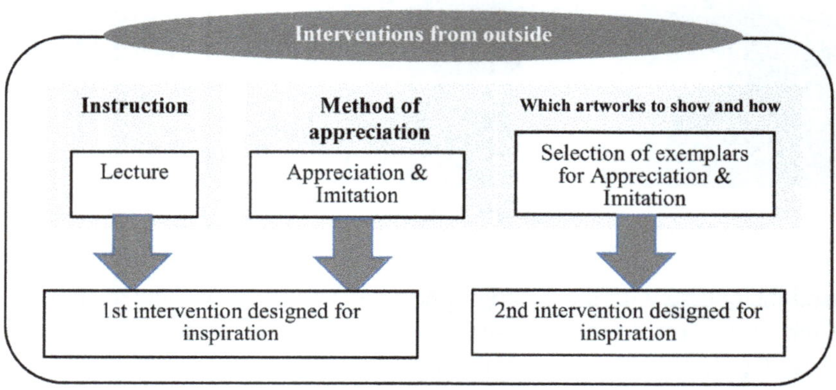

Figure 9.4 Course design according to factors to promote inspiration through others' artworks.

with an artwork, based on factors of the method of appreciating and exhibiting artworks, and determined what kind of artworks they encountered. Specifically, the students appreciated a photograph selected by the instructor of the class and shared their comments with one another. Then the instructor provided them with explanations about the photo, such as the artist's intentions, method of expression, and the historical background. After that, they imitated the method of expression of the photograph when taking their own photos. This intervention is a method of appreciation with physical activities, based on the copying method, which was shown by Okada & Ishibashi (2016) to be useful for promoting inspiration.

Considering the students' prior experience, we selected two photographs as examples to appreciate and imitate. A classic photo was used for the first appreciation and imitation class, and a street photo was used for the second class. The classic photo was 'Farm Girl' in 'People of the 20th Century' by August Sander (1874–1964), which is a collection of portrait photographs of people in the early 20th century. The street photo was 'Los Angeles, California' by Garry Winogrand (1928–1984), which is a photograph of people, on a city street, that demonstrates the social problems hidden in everyday life. 'Farm Girl' is a photograph taken using camera techniques usually used in a studio setting, which were familiar to the students after having taken the lectures. In contrast, 'Los Angeles, California' is a photograph of a natural setting, without using the camera techniques taught in the course, and was assumed to be unfamiliar to the students, even after having taken the lectures.

Besides the lecture and the appreciation and imitation sessions, the course also included free photography classes in which the students freely created their own photo artworks and a presentation in which students showed other students and the instructor five photographs that they selected from those they had taken during the course, and received comments on them. These classes were designed to provide the students with opportunities to apply the knowledge and techniques learned in the lectures and appreciation and imitation sessions and create their own artistic photographs.

Further, we gave the students two kinds of homework in order to encourage reflection on their creative activity after each class. The first homework assignment was to explain the photographs they had taken during the classes. The second assignment was to describe what they had considered and noticed about artistic creation each week, both inside and outside of the classes. These homework assignments were submitted by e-mail and through an online system.

9.4.1 Changes in the Students with Each Educational Intervention

The following sections describe how the students' knowledge of attitudes and art-making developed and how their appreciation of others' artworks changed throughout each educational intervention on the course.

First, we described how the students obtained a basic knowledge of artistic expression through the first intervention, the lecture. The students' knowledge and attitudes towards art-making were measured during the course and analysed quantitatively to examine how they changed throughout the course. Second, we examined whether or not the students' interpretations of artworks changed and became dual-focused. We also examined whether they compared others' and their own art-making through the second intervention, the 'appreciation and imitation' stage. Finally, the students presented their own artistic photos and appreciated the photos by other students in the presentation session. We hoped to discover how the students reflected on their own artistic activities during the course, and whether they continued their photographic activities after the course was over. To do so, we qualitatively analysed the students' comments on their interpretation of others' photographs and students' statements about their own artistic activities from interviews one year after the course ended. Although most of the research studies on art education have relied solely on qualitative analyses, we combined both quantitative and qualitative analyses, which enabled us to better, understand the overall effect of the course as well as the way of thinking within each student more specifically.

9.4.2 The 1st Intervention

First, we measured the students' expressive awareness (Ishiguro & Okada, 2012), which is their knowledge about and their attitude towards creative activities especially for art-making and thought to be a part of creative fluency in art-making. Expressive awareness is thought to be a part of creative fluency in art-making. Expressive awareness means having the intention to search for a match between images and ideas and a method of expression in one's creative activity. We had discovered that beginners acquired this expressive awareness through several months of photography practice (Ishiguro & Okada, 2012). Therefore, we measured the change in the students' expressive awareness during the course by using the psychological scale that we had developed (Ishiguro & Okada 2016). There are 4 items on this scale: 'When taking photos, I consider effective methods to express my images and ideas,' 'When viewing photos, I interpret how the photographers'

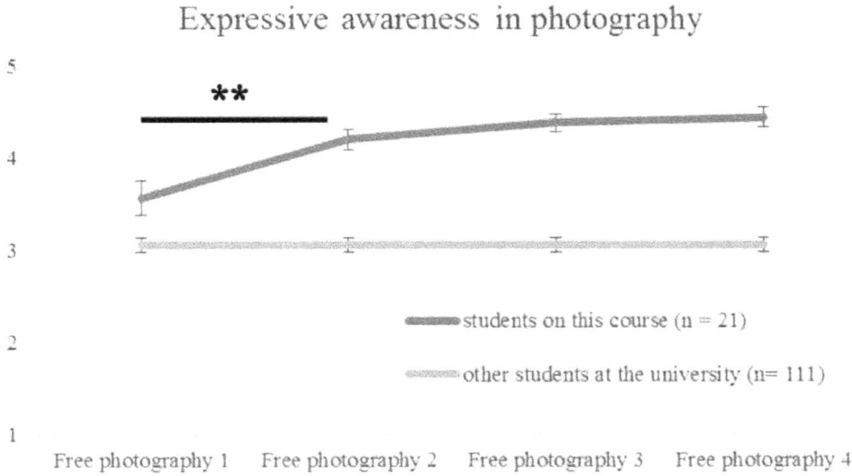

Figure 9.5 Changes in the students' expressive awareness in photography.

N.B. **:$p < .01$.

images and ideas are expressed in them,' 'Photography is a medium to express our ideas and feelings,' and 'I can improve my art-making by viewing photos by other photographers.' The results indicated that the students obtained higher expressive awareness after the lectures and retained it until the end of the course (see Figure 9.5).

9.4.3 The 2nd Intervention

In each of the two appreciation and imitation sessions, the students were provided a copy of an exemplar photograph which they appreciated for a few minutes. Students also filled in a comment sheet in which they rated the 'value' of the photo on a 5-point Likert Scale and described the reasons for their rating and what they felt and noticed about the photograph. After they had shared their comments on the photo with the other students, the instructor provided an explanation of the photo, including information about the photographer and his intentions, the expressive techniques, and the historical background of the photograph. After the explanation, the students left the classroom to take photos on the university campus, with the intention of imitating the exemplar photograph. Finally, they described what they had felt and noticed, and how their ideas about the exemplar photograph had changed after listening to the explanations by the instructor and comments by the other students. We analysed whether the students' comments on each exemplar photograph changed before and after the interventions, i.e., by sharing comments with the other students, listening to the explanation by the instructor, and imitating the photograph. We considered that the effects of the interventions might be seen not only in the answers on the comment sheets, but also in the students' homework responses after each class. Therefore, we included as data the comments that students made on each exemplar photograph in the homework they completed as part of the appreciation and imitation sessions.

In the analysis, we checked whether each student mentioned the contents in the following categories 'evaluation of the others' artworks,' 'comparison between the others' art-making and the students' own art-making,' and 'improvement in the students' own art-making'. These categories were generated according to the processes in Phase 1 (evaluation of others' artworks), Phase 2 (comparisons of the evaluation of others' artworks and reflection on the viewer's own art-making), and Phase 3 (the state of being inspired in art-making) of the ITA model. Then, we counted the number of students who described the contents in each category.

The results show that after the intervention, more students compared the exemplar photograph with their own art-making and considered the improvement of their own art-making (see Table 9.3). These results demonstrate that the intervention in the appreciation and imitation session promoted the students' dual focus and their motivation to improve their own art-making.

In addition, the changes caused by the interventions were larger in appreciation and imitation session 2 than in appreciation and imitation session 1. This difference in the students' change of interpretation between the two sessions may have been caused by the familiarity of the exemplar photographs. The photograph in appreciation and imitation 1 session was familiar and easy for the students to understand, because they had learned about the expressive technique in the lecture. Some students mentioned a comparison between the work and their own art-making and improvement of their own art-making even before the intervention. In contrast, the photograph in appreciation and imitation session 2 was unfamiliar and difficult for them to connect with their own art-making before the intervention.

9.4.4 Presentation

The students chose 5 photographs for the presentation. During the class, the students commented on each of the other students' photographs displayed on the table. Then, each student explained their own photographs and received

Table 9.3 Changes in the students' interpretations before and after the interventions in the appreciation and imitation classes and after the presentation class

	Appreciation & Imitation 1 ($n = 21$)		Appreciation & Imitation 2 ($n = 19$)		
	Before the Interventions	After the Interventions	Before the Interventions	After the Interventions	Presentation ($n = 21$)
Evaluation of others' artworks	21 (100)	20 (95)	19 (100)	19 (100)	15 (71)
Comparison between others' art-making and the students' own art-making	5 (24)	9 (43)	0 (0)	8 (42)	8 (38)
Development of the students' own art-making	8 (38)	11 (52)	2 (11)	8 (42)	9 (43)

N.B. The numbers in brackets show the percentage of students who described the contents of each category in their comment sheet and diary homework in each class.

comments from the instructor. In addition, (s)he received comments from the other students by e-mail after the class. We also analysed the students' homework responses after the presentation class and examined whether the students became dual-focused and motivated to improve their own artistic expression by appreciating the other students' photographic expressions. Table 9.3 shows that 38% of the students compared others' art-making with their own art-making, and 43% of the students considered the development of their own art-making after the presentation. Considering that the presentation session did not directly guide students carefully to appreciate the artworks of others, we could say that they became dual-focused and motivated for their own art-making by themselves through viewing others' artworks and listening to their comments.

It is important to note that the presentation was effective for the students in order to find their expression and gain motivation to pursue their originality. The following statements are some of the students' descriptions in their diary homework after the presentation class.

"Photography allows us to create artworks in any circumstances, which is different from painting. So I want to make use of the circumstances for my expression. I want to enjoy the present moment more, because I tend to rely on making plans."

"When I viewed a unique photograph by another student, I found that we can make use of our own senses, which are different from those of others. I want to try to take a more unique photo because I am usually captured by a kind of stereotype in photography,"

These statements imply that the presentation provided the students a chance to find a variety of artistic expressions by comparing the art-making of the other students with their own art-making. The comparison was effective even when the students were all beginners in artistic photo expression, were following the same course, and took photos in the same environment. Actually, the fact that there were a variety of photographic expressions among the students even in such uniform conditions may have given the students the chance to notice one another's originality of expression.

9.4.5 Changes in Students' Artistic Activities throughout the Course

The goal of the course was cultivation of creative fluency, by offering the students an authentic experience of artistic creation and by encouraging the students to commit to artistic activities. We have mentioned in the previous section that the students developed their knowledge of and attitude to artistic expression and experienced inspiration by deeply interacting with the photographs by other photographers and the other students. The next question is whether the students enjoyed their artistic activities even after the course was finished. This question is important because it is related to the development of creative fluency. To answer this question, we assessed whether their commitment to artistic activities changed from before the course to one year after the course. At the beginning of the course, we asked the students to fill out questionnaires to determine the amount of the exposure they had to some artistic activities, especially photography, before the course. In addition, we interviewed them about their artistic activities one year after the course. Fourteen of the 21 students participated in the follow-up interview (7 students could not participate in the interview because they had already graduated from the university or were studying abroad). We coded their artistic activities in photography and counted the number of students who were involved in each activity (see Table 9.4). The results show that 13 students continued their photography for purposes of creating a record, and 7 students continued their photography to make their own artworks. Four students began to show their own photos on the Internet or in photo contests. One student who received a prize in a photo contest reflected on this course in the follow-up interview as follows:

Table 9.4 Changes in the students' photographic activities before and one year after the course

		Number of Students Committed to Each Activity			
		Before the Course	One Year After the Course	McNemar's Test	
Appreciation	Encountering photographs in daily life	0	1	1	
	Appreciating photographs as artworks	3	8	3.6	
Photography	Taking photos as a record	3	13	10	*
	Taking photos as artworks	0	7	7	*
Producing their own photos		0	4	4	
Supporting creative experts		0	0		

N.B. The grey part shows the activities that students experienced on the course.
The Bonferroni correction was applied to each statistical test of difference between numbers of students engaged in each activity before and after the course.

"The course provided me with a chance to learn about artistic photo expression. (...) After the course, I bought a digital single reflex camera, started to subscribe to a photography magazine and to study photography, attended talks by photographers, and entered photo contests. Now I can say that photography is my hobby."

In addition, she discussed what she learned in the course:

"There are two parts of the course that I remember vividly. First, in every class, we had time to review other students' photographs. Although we took the photographs in a similar place at a similar time, the photographs were all quite different. This helped me to see that we all had different perspectives; despite being in the same situation, we took entirely different shots. Secondly, in each class, we appreciated artworks by professional photographers. I had not studied professional photographs before, and I found that there were some patterns in photo expression, even in the photos of professional photographers. Also, I had previously thought that most professional photographers took only portraits. But taking the course allowed me to see that there were many other types of photography, such as snapshots and street photography."

Another student commented on the appreciation and imitation sessions as follows:

"The other students and I were able to take a variety of photographs despite being in such similar locations/situations. I learned not only the importance of being artistically stimulated by something, but also how to interpret such stimulation. The appreciation and imitation sessions were critical in learning this. By imitating the professional photographers, I thought about what the differences between the professional photographs and my photo were and how I could deal with these differences."

Some of the students changed their ideas about artistic expression in general.

"I learned that it is important to think about what I want to photograph, rather than how I want to photograph it. This is not limited to photographic expression. For artistic expression, we should not stick only to techniques and methods, but also consider what we want to do and why."

These results suggest that the course effectively supported students in their creative experience and commitment to artistic activities. Additionally, the interventions designed to promote inspiration were shown to be effective in encouraging novices' learning about art. Further research is needed to generalize these findings in various learning settings.

9.5 Conclusion

Inspiration has been regarded as a mysterious phenomenon and for a long time, there has been no framework for understanding the inner process of inspiration in artistic creation. In this chapter, to answer the question of how we can utilize inspiration process in art learning settings, we provided a brief review of recent psychological studies on inspiration and explained an outline of our model of inspiration for art-making through art appreciation (ITA). On the basis of our model, we have listed factors that promote inspiration by others' artworks in art educational settings. In addition, we have described an undergraduate course on artistic creation as an illustration of our art educational practice, which was designed according to the factors mentioned above. Although in this chapter we have described only one case of an artistic photography course designed to encourage inspiration, future research should lead to an increase in such practices and the assessment of the educational effects in order to offer more practical advice.

It is highly likely that the factors mentioned above would be useful in designing educational programmes in not only the fine arts but also in other creative domains, such as music, drama, literature, or even science. In the domains of science and literature, scientists or writers often read research articles or works by others in order to learn from them and produce new works when inspired by them. Our framework for the use of inspiration in educational settings would also be useful in facilitating creative activities of this kind. Future studies are needed to examine the possibilities of applying our framework to other creative domains. Through such studies, it should become possible for everyone to experience and make use of inspiration in various creative domains.

Acknowlegdements

We are thankful to Fumimasa Hosokawa, who provided expertise in conducting the art course. This research was supported by the Grant-in Aid for Scientific Research #15H01988, Grant-in-Aid for JSPS Research Fellow #2311149, and the Ishibashi Foundation.

References

Agata, T., and Okada, T. (2013). Souzou no syutaisya tositeno shimin wo hagukumu: Souzouteki kyouyou wo ikuseisuru igi to sonohouhou (The importance of cultivating people's creative literacy). *Jpn. Cogn. Sci. Soc.* 20, 27–45.

Bullot, N. J., and Reber, R. (2013). The artful mind meets art history: Toward a psycho-historical framework for the science of art appreciation. *Behav. Brain Sci.* 36, 123–137.

Chan, J., Dow, S. P., and Schunn, C. D. (2015). Do the best design ideas (really) come from conceptually distant sources of inspiration? *Des. Stud.* 36, 31–58.

Chan, J., Fu, K., Schunn, C. D., Cagan, J., Wood, K., and Kotovsky, K. (2011). On the benefits and pitfalls of analogies for innovative design: Ideation performance based on analogical distance, commonness, and modality of examples. *J. Mech. Des.* 133, 081004-1-11.

Chemi, T., Jensen, J., and Hersted, L. (2015). *Behind the Scenes of Artistic Creativity. Creating, Learning and Organising.* New York, NY: Peter Lange.

Csikszentmihalyi, M. (1999). "Implications of a systems perspective for the study of creativity," in *Handbook of Creativity,* Ed. R. J. Sternberg (New York, NY: Cambridge University Press), 313–335.

Fu, K., Chan, J., Cagan, J., Kotovsky, K., Schunn, C., and Wood, K. (2013). The meaning of "near" and "far": the impact of structuring design databases and the effect of distance of analogy on design output. *J. Mech. Des.* 135, 021007-1-12.

Gentner, D., and Markman, A. B. (1997). Structure mapping in analogy and similarity. *Am. Psychol.* 52, 45–56.

Gentner, D. (1983). Structure-mapping: a theoretical framework for analogy. *Cogn. Sci.* 7, 155–170.

Gick, M. L., and Holyoak, K. J. (1980). Analogical problem solving. *Cogn. Psychol.* 12, 306–355.

Holyoak, K. J., and Koh, K. (1987). Surface and structural similarity in analogical transfer. *Mem. Cogn.* 15, 332–340.

Holyoak, K. J., and Thagard, P. (1996). *Mental Leaps: Analogy in Creative Thought*. Cambridge, MA: MIT Press.

Huang, Y. C., and Han, S. R. (2014). "An immersive virtual reality museum via second life: extending art appreciation from 2D to 3D," in *International Conference on Human-Computer Interaction,* eds Y. C. Huang, and S. R. Han, (Berlin: Springer International Publishing), 579–584.

Ishiguro, C., and Okada, T. (2012). "Emergence of control in artistic expressions and the process of expertise," in *Proceedings of the 34th Annual Conference of the Cognitive Science Society,* Sapporo, 1733–1738.

Ishiguro, C., and Okada, T. (2015). "The effects of art experience, competence in artistic creation, and methods of appreciation on artistic inspiration," in *Poster presented in the 31th International Congress of Psychology,* Yokohama.

Ishiguro, C., and Okada, T. (2016). Souzouteki kyouyou wo hagukumu geizyutu kyouiku zissen: Nichizyou no syashin sousaku katsudou ni oyobosu kouka (Development of creative fluency in an artistic photography course). *Jpn. Cogn. Sci. Soc.* 23, 221–236.

Ishiguro, C., Yokosawa, K., and Okada, T. (2016). Eye movements during art appreciation by students taking a photo creation course. *Front. Psychol.* 7:1074. doi: 10.3389/fpsyg.2016.01074

Kiyokawa, S., Izawa, T., and Ueda, K. (2006). "Effects of role exchange between task-doing and observing others on insight problem solving," in *Proceedings of the 28th Annual Conference of the Cognitive Science Society,* Vancouver, (1617–1622).

Langer, E., Pirson, M., and Delizonna, L. (2010). The mindlessness of social comparisons. *Psychol. Aesthet. Creat. Arts*, 4, 68–74.

Leder, H., Belke, B., Oeberst, A. and Augustin, D. (2004). A model of aesthetic appreciation and aesthetic judgments. *Br. J. Psychol.* 95, 489–508. doi: 10.1348/0007126042369811

Nakano, Y., and Okada, T. (2016). "Shokuhatsu suru communication to museum (Inspiring communication and museums)," in *Shokuhatsu Suru Museum: Bunka Teki Koukyou Kuukan no Aratana Kanousei wo Motomete (Inspiring Museums: an Inquiry into New Possibilities for Cultural Public Space)* eds K. Nakakoji, H. Shindo, Y. Yamamoto, and T. Okada (Kyoto: Airi Shuppan).

Okada, T., and Ishibashi, K. (2016). Imitation, inspiration, and creation: cognitive process of creative drawing by copying others' artworks. *Cogn. Sci.* 41, 1804–1837. doi: 10.1111/cogs.12442

Okada, T. (2016). "Shokuhatsu suru communication to museum (Inspiring communication and the museum)," in *Shokuhatsu Suru Museum: Bunka Teki Koukyou Kuukan no Aratana Kanousei wo Motomete (Inspiring the Museum: an Inquiry into New Possibilities of Cultural Public Space)*, eds K. Nakakoji, H. Shindo, Y., Yamamoto, and T. Okada, Kyoto: Airi Shuppan.

Oleynick, V. C., Thrash, T. M., LeFew, M. C., Moldovan, E. G., and Kieffaber, P. D. (2014). The scientific study of inspiration in the creative process: challenges and opportunities. *Front. Hum. Neurosci.* 8:436. doi: 10.3389/fnhum.2014.00436

Pelowski, M., Markey, P. S., Lauring, J. O., and Leder, H. (2016). Visualizing the impact of art: An update and comparison of current psychological models of art experience. *Front. Hum. Neurosci.* 10:160. doi: 10.3389/fnhum.2016.00160

Poze, T. (1983). Analogical connections: The essence of creativity. *J. Creat. Behav.* 17, 240–258.

Takagi, K., Kawase, A., Yokochi, S., and Okada, T. (2015). "Formation of an art concept: a case study using quantitative analysis of a contemporary artist's interview data," in *Proceedings of the 37th Annual Conference of the Cognitive Science Society,* Pasadena, CA, 2332–2337.

Thrash, T. M. and Elliot, A. J. (2003). Inspiration as a psychological construct. *J. Pers. Soc. Psychol.* 84, 871–889.

Thrash, T. M., and Elliot, A. J. (2004). Inspiration: core characteristics, component processes, antecedents, and function. *J. Pers. Soc. Psychol.* 87, 957–973.

Tyler, C. W., and Likova, L. T. (2012). The role of the visual arts in enhancing the learning process. *Front. Hum. Neurosci.* 6:8. doi: 10.3389/fnhum.2012.00008

Ward, T. B. (1998). "Analogical distance and purpose in creative thought: Mental leaps versus mental hops," in *Advances in Analogy Research: Integration of Theory and Data from the Cognitive, Computational, and Neural Sciences* eds K. J. Holyoak, D. Gentner, and B. Kokinov (Sofia: New Bulgarian University), 221–230.

Weisberg, R. W. (2006). *Creativity: Understanding Innovation in Problem Solving, Science, Invention, and the Arts*. Hoboken, NJ: John Wiley & Sons.

10

What Are the Enabling and What Are the Constraining Aspects of the Subject of Drama in Icelandic Compulsory Education?

Rannveig Björk Thorkelsdóttir

School of Education, University of Iceland, Reykjavík, Iceland

Abstract

The aim of this chapter is to look at what aspects enable and what aspects constrain the subject of drama in Icelandic compulsory education, using the lens of practice architectures theory. The chapter is based on my PhD study entitled *Understanding Drama Teaching in Compulsory Education in Iceland: A Micro-ethnographic Study of the Practices of Two Drama Teachers.* Based on a socio-cultural frame of understanding, an ethnographic study of the culture and the context for the implementation of drama was carried out. The ethnographic account is based on thick descriptions and thematic narrative analyses summed up as a cultural portrait of the drama teaching practices in two primary education schools in Iceland. The theory of practice architectures, proposed by Stephen Kemmis and Peter Grootenboer, was used to interpret the findings. Enabling and constraining arrangements in the practice architectures connected to the implementation of drama as a subject in compulsory education. The findings reveal that the enabling aspects of drama teaching are less visible than factors that constrain the teaching.

10.1 Introduction

There is something special about the art form of drama and how it can work, as a practice aimed at learning in general, and as a subject on its own. In Icelandic school, drama is presented in the curriculum both as a subject and as a method. Through drama, the students can learn to interact with one another in a safe space, try out different societal roles. Role-playing offers students the opportunity to explore aspects of what it means to be human. This chapter is based on my PhD study entitled Understanding Drama Teaching in Compulsory Education in Iceland: A Micro-ethnographic Study of the Practices of Two Drama Teachers. The research project was motivated by the fact that drama was included in the Icelandic curricula in 2013 as compulsory subject for all students in primary and lower secondary schools. To include a new subject in the curriculum raises many questions regarding how drama can contribute to students' learning within the arts. What can be learned by taking drama? Who has the competence to teach this subject? And if it is used as a method to support learning in other subjects, who can elaborate on this working form in a way that best brings out its potential as an art subject? When a new national curriculum guide for drama is created, will there arise a need for continuous education to meet the changes introduced? How can teachers become qualified to teach drama, that is, those who may not be used to including drama in their teaching? What kind of support is needed from the education system in order to make this work? What can this particular arts subject contribute that may not be as easily accomplished in other subjects? Ultimately, what could enable or constrain the drama teaching practices? In this chapter, the main focus will be directed toward how *the enabling arrangements and the constraining arrangements are manifested in the practice architectures of the subject of drama within Icelandic compulsory education.*

10.2 Learning Through the Arts

The arts make a strong claim as parts of the education system. Through the arts, students can construct new aesthetic knowledge and deepen their human impulses and experiences. Drama is by nature an integrative practice where all art forms are combined. According to Mike Fleming, "Arts enrich our understanding of the world, challenge prevailing ideologies, widen our perspectives, engage and delight us, and celebrates out humanity", (2012, p. 1). In his The Arts in Education, An Introduction to Aesthetics, Theory and Pedagogy, Fleming writes about learning in and through the arts. He says:

> Learning through the arts looks beyond the art form itself to outcomes that are extrinsic and often take place when arts are employed across the curriculum to further learning in other subjects. Learning in the arts more often refers to learning within the

> subject itself /.../ however it is when the concepts become less distinct and start to merge that the greater interest and insight is found. (p. 68)

Fleming also claims that teaching art must involve more than simply teaching children to express themselves through creating art.

> Teaching children to appreciate art must involve due attention to both the art object and their experience in relation to it. With regard to content it is reasonable to suggest that students should be taught to participate in the cultural world in which they will live with its diverse range of forms and types of art. (p. 45)

10.2.1 Learning Through Drama

Michael Anderson (2012) points out that drama sits in a unique place within the education system, at the intersection between intellectual, creative and embodied education. Furthermore, Anderson holds that drama teaching is transformative, meaning that drama can support the academic, social and emotional growth of young people. Drama education, and arts education in general, is a pedagogy with a heritage that has the potential to modernize schooling (p. 10). According to John O'Toole and Joanne O'Mara (2007, p. 207), there are four "paradigms of purpose" attached to the use and teaching of drama. They are: *cognitive/procedural*, which means gaining knowledge and skill in drama; *expressive/developmental*, which means growing through drama; *social/pedagogical*, which means learning through drama; and *functional/learning*, which means learning what, people do in drama. However, in many texts about drama in education these purposes are interwoven. There is also a bigger picture that drama education as cultural activity fits into. Jonothan Neelands (1996, p. 29) described modes of empowerment in drama on different levels, from personal, cultural, communal to social/political, as shown in Table 10.1.

In Table 10.1, theatre is described as a personally transforming resource, as a means of making the invisible influences of culture visible, and as an act of a community, which mirrors its hopes, fears and dreams. Finally, theatre can be seen as a rehearsal for political change and as an arena for radical dialogue. These basic values in theatre can underpin drama both as an art form and as pedagogy. Neeland's list shows the empowering potential of theatre, which ultimately acknowledges the role that drama and theatre education, plays in increasingly multicultural and diverse settings.

Table 10.1 Modes of 'empowerment' in drama according to Neelands (1996, p. 29)

Personal	Theatre as a personal transforming cultural resource: Through using and engaging with theatre one's sense of 'self is transformed; learning about genres, histories and the range of "choices" of form is part of personal empowerment through theatre.
Cultural	Theatre as means of making the invisible influences of culture visible and discussable; theatre as a mirror of how we are made; theatre as a mirror of who we might become.
Communal	Theatre as an act of community in which we actively participate in making of communal representations; theatre as social and aesthetic expression of a community's hopes, fears and dreams.
Social/Political	Theatre as rehearsal for change and as an arena for radical dialogue.

10.2.2 Drama in the Icelandic National Curriculum

In the Icelandic National Curriculum Guide (2014) for compulsory education (grades 1–10), arts and crafts are divided on the one hand into performing arts (dance and dramatic arts), visual arts and music, and crafts, on the other, which includes home economics, design, and craft and textiles. The timetable for arts and crafts should account for around 15% of the weekly classes. Each compulsory school then decides whether the subject areas should be taught separately or be integrated (Österlind et al., 2016). According to the curriculum, drama education should train students in the methods of the art form, but no less in dramatic literacy in the widest sense of the term, that is, by enriching and facilitating the students' understanding of themselves, human nature and society.

> In drama students are to have the opportunity to put themselves in the position of others and experiment with different expression forms, behaviour and solutions in a secure school environment. Drama encourages students to express, form and present their ideas and feelings. In addition, drama constantly tests cooperation, relationships, creativity, language, expression, critical thinking, physical exertion and voice projection. This is all done through play and creation. (Ministry of Education, Science and Culture, 2014, p. 153)

The competence criteria for drama in the Icelandic National Curriculum Guide also provide aims for drama as a teaching method, where the teaching methods are grounded on the art form itself. The competence criteria for drama in grade 4 are process-based, whereas for grade 7 and 10, lessons in drama are theatre-based, and drama aims towards the product field. Drama is considered a subject which relies on active participation, for example making a play, performing a play and responding to a play (Thorkelsdóttir, 2016).

10.3 Theoretical Perspective

The research question for this chapter is: *What are the enabling aspects and what are the constraining aspects of the subject of drama in Icelandic compulsory education*? The chapter is based on my PhD study which brings to bear the practice theory of Stephen Kemmis and Peter Grootenboer (2008), especially the theory of practice architectures, which serves as the overarching educational theory. Practice theory is a family of theoretical perspectives that are part of a practical turn, where the aim has been to understand different practices, focusing on the knowledge in a practice (Østern 2016, p. 21). *The theory of practice architectures,* formed by Kemmis and Grootenboer (2008), falls under a particular practice theory about education. This theory builds on Theodore Schatzki's practice theory (1996; 2002), in which a practice is defined as a nexus of *sayings* and *doings*. Moreover, Kemmis, Jane Wilkinson, Christine Edwards-Groves, Ian Hardy, Grootenboer and Laurette Bristol (2014, p. 31) have added '*relatings*' to their definitions of practice. They describe practice as organized bundles of sayings, doings and relatings, which hang together in a project of practice, where that practice itself figures the overall purpose that gives it coherence. Kemmis et al. (2014) also maintain that education needs to be in a continuous process of change and transformation. In this respect, a practice theory is critical, but they emphasize that this continuous transformation needs to be undertaken and initiated from those inside the practice rather than external agents.

10.3.1 The Practice Architectures

The concept *practice architectures* refer to the specific cultural-discursive, material-economic and social-political arrangements found in or brought to a site (Kemmis et al., 2014). The term practice architectures can evoke associations of a building or something that is grounded. Architectures, as Kemmis et al. present them, are invisible, yet detectable, social patterns or arrangements formed in different cultures of practice that anyone, that wants to take part in that practice, has to operate under. When a person enters a practice, there are already practice architectures in place that regulate what can be said and how, what can be done and how, as well as determining how relations, hierarchies and solidarity function. These arrangements can both enable and constrain a practice. In other words, these are arrangements that decide whether the practice is possible or not (Thorkelsdóttir, 2016). As mentioned above, Kemmis et al. (2014) explain practice as organized bundles of *sayings*, *doings* and *relatings* that hang together in the project of practice. This is about what is considered appropriate to say or do in a particular practice, and what kinds of relations between people, within the practice, are viewed as necessary and proper according to the culture of that practice. Practice is socially established cooperative human activity in which people become speakers of shared languages and develop shared forms of understanding. They also take part in activities (doings: what people describe as skills and capabilities) and they share ways of relating to each other (relatings). Simply put, the project of practice is what people answer when asked: "What are you doing?" while they are engaged in the practice. In *Changing Practices, Changing Education,* Kemmis et al. (2014) developed this theme into a theory of education, related to the theory of practice architectures, shown in Figure 10.1.

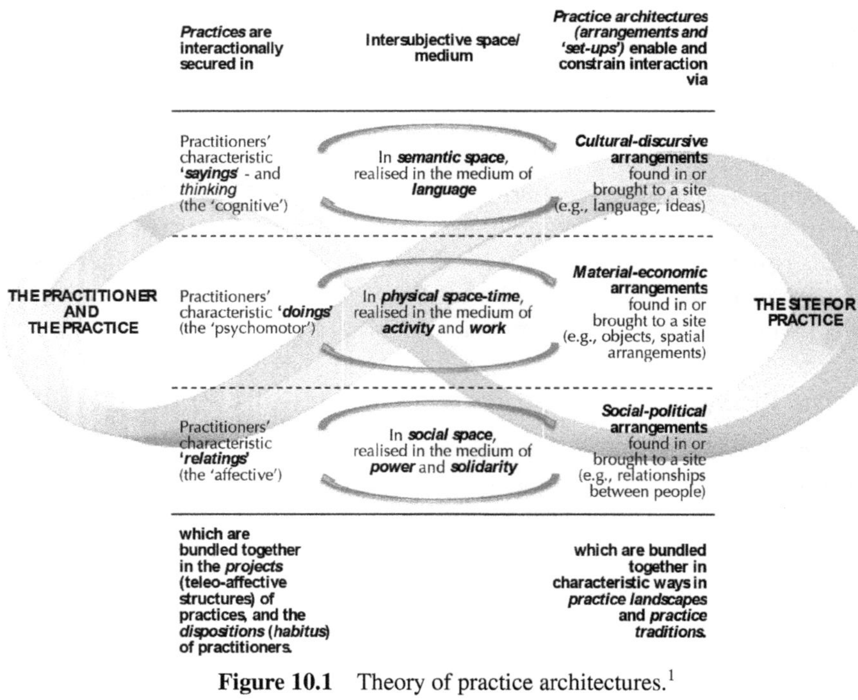

Figure 10.1 Theory of practice architectures.[1]

[1]Kemmis et al. (2014, p. 27; with permission from Kemmis et al.).

In Figure 10.1, three columns are displayed: To the left, the practice and the practitioner with sayings, doings and relatings; to the right, the arrangements in the practice architectures consisting of cultural-discursive, material-economic and social-political arrangements. The column between the practice column and the practice architecture column is called intersubjective space, where communication between the practice and the practice architectures is realized in the semantic space realized through the medium of language, physical space-time realized in the medium of activity and work, and social space realized in the medium of power and solidarity. Figure 10.1 also displays a "feedback loop" centred in intersubjective space. The feedback loop can be characterized as a 'learning loop' within intersubjective space, as a space for communication, or in other words, a communicative space (Thorkelsdóttir, 2016).

10.3.2 Feedback Loops and Dialectical Tensions

Before I explore the enabling and constraining aspects of drama teaching practices, I will briefly explicate the feedback loops and their relations to dialectical tensions. The feedback loops have the potential for learning to happen, transforming the understanding of the participants. The dialectical tensions can be of different kinds, but one main tension is the one that arises between the personal dreams, visions and hopes, on the one side, and 'the reality' of the practice architectures, on the other.

The tensions generally exit between some specific practice and a set of constraining or enabling factors within the practice architectures. The theory of practice architectures contends that when a teacher enters a community of practice, like a school, this site is in part already formed (Thorkelsdóttir, 2016). Practice architectures are made, in part, by the previous practices of people in the site, yet these architectures are not entirely rigid and can be changed by people's practices.

The practice architectures of a specific site can be identified in the sayings (cultural-discursive arrangements), doings (material-economic arrangements), and relatings (social-political arrangements). Kemmis et al. (2014) argue that changing education involves not just changing the way teachers teach or students learn. Rather, changing education necessarily entails an alteration of the practice architectures found in the particular educational sites. Different practices co-exist in interdependent relationships with one another. Hence, by changing education, like adding a new subject such as drama into a curriculum, it is not enough just to change the way teachers teach or students learn. The implementation must also instigate change in other parts of the practice to the extent that it influences the practice architecture as a whole, not just parts of it.

10.3.3 Data

The data for this chapter is based on an ethnographic study of two drama teachers in two schools in Reykjavík, over one school year, with observations and interviews providing the main sources of information about cultural behaviour that characterized their practice. The ethnographic account is based on thick descriptions and thematic narrative analyses summed up as cultural portrait of the drama teaching practices in the Hillcrest and Mountain-line schools, respectively. Both schools had a *scheduled time for drama* over the school year 2013–2014. At Mountain-line school, drama has been part of the school curriculum for five years and it is a tradition in the school that each spring the 6th grade puts on a performance in the forest. Drama was not on the Hillcrest school timetable before 2013, however, the year before, the music teacher (2012–2013) combined drama and music in her music lessons. The main method by which I have analyzed my material is interpretative. I have searched for meaning and understanding and moved between the various parts and the whole in order to develop an understanding of the emerging culture in the drama class.

10.4 Four Perspectives on the Implementation of a Drama Curriculum

The cultural portrait of the implementation of drama was interpreted from four perspectives, embracing discursive constraints as well as discursive opportunities when teaching drama. The four perspectives are: (1) The researcher perspective: The researcher's observation of the drama classroom teaching practices; (2) the drama teacher perspective: The narratives of the drama teachers and their learning trajectories; (3) the student perspective: The students' experiences of drama; and finally (4) the principal perspective: The principals' perspectives of drama in their schools.

The drama class was the unit of analysis. The four perspectives elaborated upon can be said to represent both insider and outsider perspectives. On the one hand, the insider perspectives are extracted from the teachers and the students. On the other, the researcher and the principals represent outsider perspectives, at least to a larger degree than the persons directly involved in the practice of drama.

Table 10.2 Complexity of the drama teacher role and drama teaching relating to the implementation of drama

The Complexity of the Drama Teaching			
Researcher Perspective	Teacher Perspective	Student Perspective	Principal Perspective
The learning culture The playing culture The importance of experience The importance of collaboration, The complex teacher role	Recognition as a drama teacher Professional identity Belonging to a community Struggling with the pedagogy of drama teaching	Have fun, perform and learn To learn to create a character, use props, light, music and costumes To learn to work with everybody even if you do not want to To learn to listen to each other and to play out a story The drama teacher is important	Importance of the performance The students have fun while they learn about culture. Drama is promoted as a brand

The drama teacher's work as presented in the literature	
A paradigm of purpose consisting of	Drama's potential for empowerment on
making/forming/creating, presenting/performing/ communicating, responding/reflecting/appraising. (O'Toole & O'Mara)	personal level, cultural level, communal level, social/political level. (Neelands)

In Table 10.2, I have visualized the complex expectations that the drama teachers meet in the everyday life of the school, such as making drama and learning in drama through games that aim at performance. I concentrated on how the literature describes the "ideal" content of the practice, the empowering potential, and the qualities of a drama teacher and the importance of professional identity. Juxtaposed with this is the teacher's need for recognition and community that that ties directly with intersubjective space (Kemmis et al., 2014) in the drama practice.

10.5 The Intersubjective Spaces in the Drama Teaching Practice

I will now elaborate further on the three intersubjective spaces that characterize the drama teaching practice and that have a mediating function between practice and practice architectures. These are the semantic space mediated by language, the time space mediated by activity, and the social space mediated by relations, power and solidarity. After this, I interpret and summarize the enabling and constraining feedback loops connected to the drama teaching practice.

10.5.1 The Importance of the Semantic Space

The semantic space is ideally a communicative space where the actors in the field talk and think about and elaborate on their plans for practice. In the context of the Hillcrest and Mountain-line schools, the "actors" in the semantic space are the drama teachers in the two schools, the students, the other teachers at school, the principal and the parents. The cultural-discursive conditions in the practice architectures contribute in site-specific ways to what can be talked about and what can become the subject of negotiation. In the actual study, I have used the terms "emerging culture for drama" to describe this process, because the culture of drama, as an arts subject, is not yet firmly established. Drama is a newcomer in the curriculum.

One of the two drama teachers claimed a space for drama and managed to obtain it over the course of a few years. The other teacher did so as well, but was not as successful in the period of the school year 2013–2014. However, the *culture of performance* seems to have been established in both schools. In fact, it can be claimed that the performance holds the drama practice "in place", as suggested by Kemmis and Grootenboer, with regard to the role of the practice architecture.

In Hillcrest school, the *learning in drama is considered fun because of the play and games*. The students enjoy the drama class and they think that the class is fun. The school had adopted drama as a subject, that is not in conjunction with music, for the first time during the school year 2013–2014. The principal of the school expressed that the school should take part in the *cultural upbringing of the students in the community,* and that there should generally be more creativity in all the classes. In spite of these aspects, what constrains the drama teaching in the school is more visible than what enables the teaching. What stands out as the central constraining factor is the workload, due mainly to many mandatory tasks. The method of communication, or the lack of it within the school, leads to stressful situations, as one of the teachers did not claim the space needed among other teachers, thus limiting her space for constructive social interactions with others.

In Mountain-line school, *creativity is one of the values* of the school and it is also one of the fundamental pillars in compulsory education according to the national curriculum. The school emphasizes arts and crafts and the *principal supports drama* and claims that he would like to develop it further in his school. *Drama as a production* is important in the school, especially the traditional forest play that has influenced *creativity in the school culture.* The students like the drama classes and drama is considered fun. The bundle of sayings, doings and relatings in one of the schools, as well as the discursive, material and social arrangements, can be seen to be more supportive to drama than in the other school, as they appear in time and space over one school year.

10.5.2 The Importance of Activity in Space-Time

The activities carried out in the drama class are dependent upon the collaboration and communication between the teacher and the students. This is, of course, of vital importance, but here the material-economic arrangements interfere strongly with what kind of activities can be carried out, and for how long. For example, if teachers come in to the classroom in the morning and find that other teachers left the desks and chairs spread out on the floor at the end of yesterday, this can be considered a constraining aspect within the practice architectures. Through the practice of re-arranging the furniture, the teachers can change the way the site is currently structured. Thus, the teacher is able to change the practice architectures in ways more suitable for drama teaching. The purpose spelled out in the national curriculum guide for drama should be more effectively realized in the actual work in the classroom.

Drama is one of the arts subjects, and it should be taught to all students, which might be challenging for many teachers, especially when they are trying to maintain leadership of a big group with varying motivations. Within the practice architectures, the priorities of both schools are visible in that they both have defined positions for drama teachers and also special rooms designated for drama. The activities brought into the drama class by a skilled teacher contribute to cultural learning, and learning about the art form, such as learning how to make a script, how to improvise, how to build a character and how to think about the dramatic process and its structure. The physical space-time makes the practice possible. Mountain-line school *has good facilities for the arts* and crafts and the arts subjects are all taught in the same corridor, each with their own classroom, which enables the drama teaching practice. In Hillcrest school, the fact that the drama-teaching practice takes place in a classroom that is used and designed for music teaching is a constraining factor.

10.5.3 The Importance of the Social Space: Relations, Solidarity and Power

The ways in which the actors (the drama teachers, the students, the other teachers at school, the principal and the parents) relate to each other in the intersubjective space seems to be of utmost importance for the drama teachers. The teachers' collaboration and the adjustments of the teaching practices were favorable in Mountain-line school.

The drama teacher was given fewer mandatory activities in May when he was working on the performance of the forest play, and he had more control of his time. The material and economic arrangements in the school supported the work with the forest play and prioritized drama in general creating constructive practice architecture for drama. The teachers received extra payment for their work on the forest play and money was put into costumes, props and transportation. The fact that all the arts and crafts teachers have their own classroom in a special corridor makes the collaboration easier for the teachers, thus enhancing the social feedback loops. The forest play attracts a lot of media attention for the school, which is considered to be one of the leading schools with regard to drama teaching on the elementary school level in Iceland.

What could constrain drama teaching in the future? For example, if drama were taught in smaller groups (not as a whole class), which would no longer make it affordable. Moreover, it could lead to draining the school's budget and, in that way, influence the material-economic conditions of the school.

10.5.4 Enabling Feedback Loops of the Practice Architectures in Drama Practice

I will now summarize some of the enabling aspects for the drama teaching culture. These aspects have an impact on the enabling feedback loops between the practice and the practice architectures.

That **drama has an established place** in the national curriculum is a fundamental prerequisite in the practice architecture that enables the drama teaching practice. The status achieved should not be undervalued. At the same time, this fact represents a challenge for the educational system. One of the fundamental conditions for making the curriculum work in practice is teacher competence. A big challenge facing the implementation of drama in the curriculum has to do with whether or not there are enough teachers with the **right qualifications to teach drama.** Especially in this newcomer situation there are likely not to be enough qualified teachers. Both schools had prioritized the hiring of qualified drama teachers. This serves as an enabling factor. Principals are main actors in **enabling a culture for drama**. They are faced with the central leadership challenge of holding the practices of the school "in place", especially considering the varying arrangements present in the practice architecture. School leaders who think of arts education as valuable often find a place for these subjects in the syllabus, and promote a culture for the arts in education.

Having a **welcoming community in the school**, and being included and supported, seem to contribute to the students' perceived satisfaction of drama class and to a good feeling of recognition for the drama teacher. Thus, the school community is part of the cultural discursive arrangements as well as the socio-political arrangements. The material-economic arrangements, such as ample provision of space for drama e.g. in the form of a drama studio, or at least a room suitable for drama, can also be considered part of the goodwill of the school community, thus either contributing toward an enabling or restricting feedback loop. That drama can be organized as a **whole class activity** is a part of the school's economic arrangements and prioritizes the subject in its material-economic arrangements. What both teachers in both schools appreciate are the possibilities for **teacher collaboration.** The dialogical meeting points with the other participants are made visible through collaborations as the dialogue encourages the teachers within each school to work together. This caters to drama as a teaching method in subjects other than drama, where drama can be utilized to support the learning of other subjects.

However, it is most clearly pronounced in the preparation of performances; it qualifies the theatre performance and helps with a lot of practical arrangements. Furthermore, it is important for all subjects that the teachers consider the **subject they teach to be of importance.** As drama does not in general have a strong place in the school community, it is worth noting that the drama teachers in both schools consider their passion for the subject to be a driving force in their teaching, but they want to go deeper into the art form, and with the restricted time in school they both seem to need inspiration for their professional development outside of school in the theatre groups that they have created. Here again, the dialogical meeting points are met through collaboration with other teachers when travelling in and out of different practices. The practice architectures of the school are strengthened outside the school, and the communicative spaces are extended to response loops with a larger community.

10.5.5 Constraining Feedback Loops of the Practice Architectures in Drama Practice

I will now summarize the aspects that constrain the drama teaching culture. These constraining aspects have a negative impact on the enabling feedback loops between the practice and the practice architectures.

Both drama teachers in this study consider **heavy workload** a constraining factor with regard to the time needed to adequately prepare for and focus on complex drama teaching. Drama is a new subject in the curriculum and it has not yet claimed its space, which was clearly manifested in the fact that both teachers, who were at the beginning of their profession, had to teach other subjects in order to **obtain a full time teaching position.** Their positions as drama teachers were part time. Combining drama teaching with other subject teaching leads to increased pressure on teachers, this in turn can result in teachers becoming worn-out at an early stage of their professions. Because drama is a **whole class activity,** the schools can prioritize the subject via their material-economic arrangements. Organizing drama teaching as a whole class activity gives the teacher an opportunity to work with variably sized groups and in which several groups are sometimes combined, for example, for a performance. This can leave the drama teacher teaching up to 50 students at a time. Both teachers in both schools talk about the need for communication with other teachers, and where this lack of community creates a feeling of **isolation**. If a drama teacher does not have a fruitful relationship and dialogue with the other teachers, for example in the arts, then teaching drama is in the danger of becoming a **lonesome profession.**

As drama is new in the curriculum, the schools do **not have special drama classrooms,** similar to the ones they have for music and the fine arts. Drama is taught in a space that is designed for other subjects, and that space is often shared with music or dance. What drama needs is a classroom with drapes and lights and soft floors and storage: the bigger the room the better (Thorkelsdóttir, 2012; Sigþórsdóttir, 2009). In this research, these requirements were partly met in Mountain-line school but not in Hillcrest school. Moreover, the work of the drama teacher is often only visible in a school production. What goes on in a drama classroom can look like misaligned chaos for those who do not know what drama teaching is about, and that can lead to ignorance of the **drama teaching practice among other teachers**. The isolation of being the only drama teacher in the school and the lack of both dialogue within the school, and a dialogue with other teachers in other schools, can lead to a lack of recognition of a drama teacher's profession.

10.6 Conclusion

In this chapter, I have shed a light on the complexity of drama teaching and drama teacher role in relation to the implementation of drama in schools. The drama teaching was scrutinized through the lens of practice architectures, focusing on the enabling and constraining aspects of the drama teaching practices. What stands out, as a constraining factor, is the heavy workload that is due to many mandatory tasks. In Iceland, drama is considered a subject based on an activity that involves the making of a play, performing a play and responding to a play: making, presenting, and responding. Drama cannot be used fully as a method without being rooted in the theatre. The subject is an arts subject and the methods build on the didactic approach of the arts. If drama is to function, it is important to acknowledge its different forms, or, the different functions of drama. And perhaps now is right moment to find reasons for making drama an integral part of every teacher's repertoire – that every student teacher gets a fundamental practice in drama, knows what it is about, what it needs and where he/she can receive and could reach for assistance, collaboration and support in order to enact drama in his or her teaching practice. From the learner's perspective, it is important to articulate the learning that can be the outcome of drama and to acknowledge the ways of learning offered by drama, such as embodied and playful learning as well as deep learning. In this context, it would be fruitful to introduce multi-literacies to better gauge what students learn when producing products in drama as well as by utilizing resources on social media and the Internet.

References

Anderson, M. (2012). *Master Class in Drama Education. Transforming Teaching and Learning*. London: Continuum.

Fleming, M. (2012). *The Arts in Education. An Introduction to Aesthetics, Theory and Pedagogy*. London: David Fulton Publishers.

Goodlad, J. (1979). *Curriculum Inquiry*. New York, NY: McGraw-Hill Book Company.

Kemmis, S., and Grootenboer, P. (2008). "Situating praxis in practice," in *Enabling Praxis: Challenges for Education*, ed. S. Kemmis, T. Smith and J. Smith (Rotterdam: Sense Publishers), 37–62.

Kemmis, S., Wilkinson, J., Edwards-Groves, C., Hardy, I., Grootenboer, P., and Bristol, L. (2014). *Changing Practices, Changing Education*. Dordrecht: Springer.

Ministry of Education, Science and Culture (2014). *The Icelandic National Curriculum Guide for Compulsory Schools – with Subjects Areas.* Reykjavík: Ministry of Education, Science and Culture.

Neelands, J. (1996). "Agendas of change, renewal and difference," in *Drama, Culture and Empowerment: The IDEA Dialogues*, eds J. Toole and K. Donolan (Brisbane: Idea Publications), 20–32.

O'Toole, J., and O'Mara, J. (2007). "Proteus, the Giant at the Door: Drama and theatre in the curriculum," in *International Handbook of Research in Arts Education,* L. Bresler (Dordrecht: Springer), 203–218.

Østern, A.-L. (2016). "Betydningen av responssløyfer i veiledningsprosesser En studie av en teori om praksisarkitekturer i utdanning [The importance of response loops in supervision processes. A study of a theory about practice architectures in education]," in *Veiledningspraksiser i bevegelse (Supervision practices on the move),* eds A.-L. Østern and G. Engvik (Bergen: Fagbokforlaget), 19–44.

Schatzki, T. R. (1996). *Social Practices: A Wittgensteinian Approach to Human Activity and the Social.* New York, NY: Cambridge University Press.

Schatzki, T. R. (2002). "Introduction: practice theory," in *The Practice Turn in Contemporary Social Theory*, eds T. R. Schatzki, K. Knorr-Cetina, and E. von Savigny (London: Routledge), 10–21.

Sigþórsdóttir, Þ. (2009). *Leikur, tjáning, sköpun. Handbók fyrir leiklistarkennslu. [Play, expression, creativity. Manual for drama teaching].* Reykjavík: Námsgagnastofnu.

Thorkelsdóttir, R. B. (2012). *Leikið með listina [Play with Drama].* Reykjavík: www.leikumaflist.is.

Thorkelsdóttir, R. B. (2016). *Understanding Drama Teaching in Compulsory Education in Iceland: A Micro-ethnographic Study of the Practices of Two Drama Teachers.* Trondheim: Norwegian University of Science and Technology.

Österlind, E., Østern, A-L., and Thorkelsdóttir, R. B. (2016). Drama and theatre in a Nordic curriculum perspective – a challenged arts subject used as a learning medium in compulsory education. *Res. Drama Educ.* 21, 42–56. doi: 10.1080/13569783.2015.1126174

11

The Art of Co-Creating Arts-Based Possibility Spaces for Fostering STE(A)M Practices in Primary Education

Pamela Burnard, Tatjana Dragovic, Susanne Jasilek, James Biddulph, Luke Rolls, Aimee Durning and Kristóf Fenyvesi

Faculty of Education, University of Cambridge, 184 Hills Road, Cambridge CB2 2PH, UK

Abstract

This multi-method project asks what the arts, in transdisciplinary learning spaces, can contribute to primary education. We position our work at the interstices of: (i) the rising wave of interest in, and imperative for, building sustainable creative futures; (ii) the concept of the Anthropocene which pulls together ideas of environmental change and the relevance of education for society; and (iii) allowing pedagogical experimentation across subject boundaries to broaden learning opportunities for empowering children. Methodologically, we use the dimensions of an arts-based perceptual ecology, where direct experience, the magical (a special and exciting quality that makes something seem different from ordinary things), intuition, imagination, art-making and the language of pattern become part of an interconnected learning system, to explore how children and their teachers co-create possibility spaces for fostering transdisciplinary learning. Using a sculpture installation as stimulus, we analyse three sets of transdisciplinary practices enabled by the co-creation of possibility spaces. The first set of practices featured an artist-led workshop for teachers to engage with an art piece, to think with their hands and to work with clay as a generative event for planning curriculum spaces for transdisciplinary activities for their children. In the second set, children were invited to respond creatively to the art piece and consider the world as something we make rather than as something we inherit. The third set featured use of the language of patterns, connecting mathematics and art, or mathematics and music. We analyze and interpret the practices using an arts-based perceptual ecology (ABPE) framework from which to see how the arts, along with science, technology, engineering and mathematics (STEM), as a 'bracketed' concept (STE(A)M), is included, located and functions. We conclude and offer recommendations for the innovative work of primary schools as a whole – that is to say, the curriculum, pedagogy and the wider life of the school – to co-create the spaces for this kind of arts-based perceptual ecology in enhancing science, technology, engineering, arts and mathematics (STE(A)M). We argue for the ongoing and necessary theorizing of possibility spaces for arts positioning within STE(A)M education.

11.1 Creating Possibility Spaces for STE(A)M in Primary Education

The urgency for charting new sustainable development goals (SDGs) for the Earth's resources, on which human life depends, is irrefutable, and is being played out across local and global scales (Leach et al., 2012). Similarly, there has been a great deal written about the need for pushing out the boundaries of how we learn at all levels, from K-12 to college and graduate levels, but, particularly, starting with the possibilities in primary education. Meeting the challenges of the 21st century needs to begin in primary education. In the UK, the Warwick Commission Report (2015) reminds us that, globally, our education systems should be creative learning landscapes, infused with possibility spaces. In this scenario, important challenges arise for curriculum design and enabling pedagogical experimentation across subject boundaries to broaden learning opportunities for children. The addition of the 'A' to STEM subject teaching comes as a means of humanising science and technology-enhanced learning: for it to be informed by the arts (Ge, Ifenthaler & Spector, 2015) as well as a site for developing technological expertise (see Chung, 2007); and for it to involve active learning strategies (Fenyvesi & Lähdesmäki, 2017). Defining STE(A)M reflects an articulation of what is absent from, or problematic with, any particular author's conception of STEM education rather than a positive ascription of a subject domain or pedagogical approach.

Much of the STE(A)M literature echoes a view of the arts as valuable both intrinsically and instrumentally; the arts are deemed to be social, inclusive, humanising and thereby significant for human development in society (Belfiore & Bennett, 2007; Canatella, 2015). Arts education literature proposes that engagement with the arts in education matters in children's education (Wade-Leeuwen, 2016; Korn-Bursztyn, 2012) Specifically, in STE(A)M education, the arts intrinsically bring a roundedness to the educational experience of pupils in which they can connect different aspects of their own and other human experience and practice (Trowsdale, 2016). As such, it is argued that the arts retain their legitimacy as specific and equally valuable perspectives on the world. The addition of the 'A' also signals more creative pedagogies (Sefton-Green *et al.,* 2011) thereby giving children a more positive view of and engagement with the STEM subjects. Whilst we strongly advocate for this position, by including the (A) in brackets, we continue to problematize how the arts are configured alongside other subject disciplines in STEM education. We want to continue to ask questions to strengthen the position that arts-based practices generally, and art-making specifically, is integral to ways of being, becoming, knowing and not knowing that are experienced as radically different; all of which may be enacted in transdisciplinary forms of inquiry and learning (Marshall, 2014). Studies often characterise the use of the arts *as a way of knowing* in STEM education. There still remains, however, much debate about the integration and infusion of the arts across the primary school curriculum. How curriculum subjects can work together in primary education, rather than as separate subjects (Russell-Bowie, 2009), remains under-theorised.

Some 20 years ago a path-breaking UK creativity researcher, Anna Craft (2015), coined the term 'possibility thinking' (PT). Co-researching with teachers and practitioners in early years and primary classrooms, Craft, along with her collaborators, sought to identify the nature of PT together with pedagogical strategies that seem to foster 'what if' and 'as if' thinking in children aged 3–11. Classroom and policy strategies were devised to foster the development of PT which generates novelty operationalized by question-posing, play, immersion, innovation, being imaginative, taking risks, and self-determination within the enabling contexts of time and space which foster it (see Burnard et al., 2006). Beyond the seminal studies lead by Anna Craft and her team (Burnard et al., 2006), numerous subsequent studies were undertaken, fuelled by the potential of PT, the generative practices of artists working in collaboration with primary educators, and the importance of co-creating education futures to consider 'how' children 'might' learn creatively for the 21st century (Craft, 2015).

It could be considered that we are living in the age of the Anthropocene, a period in history where the planet has been irretrievably changed and altered by human activity and practices. Within an apocalyptic context, there is growing movement for reconciling art and science for sustainability (Stirling, 2012); that is, integrating art and science for transformative understanding and building social ecological resilience (Westley, Scheffer & Folke, 2012). There is also significant interest in transdisciplinarity: that studying Science, Technology, Engineering and Mathematics (STEM) as separate subjects is increasingly seen as an anachronistic task even for those going on the path to become scientists and mathematicians. It is concerning that children in primary education think that STEM subjects lack creativity and often do not relate to the aspirations they have for themselves (Mendick, Berge & Danielsson, 2017). The rise of the term STE(A)M, to denote the inclusion of the Arts (the (A) in STE(A)M), has the capacity to embrace and transform the landscape of 21st century learning in primary education. In this chapter we attempt to capture the essence of the potentials of subjects for synergy. Studies often characterise the impact of exposure to arts-based representation of scientific knowledge in terms of the participants' ecological awareness or the role of arts in the creation process of scientific knowledge and the scientific value of seeing everyone as an artist and experiencing aesthetic qualities that arise from perceptual appreciation in the experience of art-making (van Boeckel, 2013). We now turn to the practice of arts-based perceptual ecology as a heuristic method for learning about the language of place.

11.1.1 Introducing Art-Based Perceptual Ecology (ABPE) Methodology

American artist and scholar Lee Ann Woolery (2006) introduced the concept of *Art-Based Perceptual Ecology (ABPE)* as a way of knowing the language of place. Woolery identified six core dimensions of ABPE-methodology. (Perception is understood here as the process of making meaning out of sensation.)

The six core dimensions that comprise the art-based perceptual ecology (ABPE) framework which we applied as tools to conduct thematic analysis, with multiple readings of each practice, are:

i. **Direct experience** which relates to recognition of the role of the body as the connection between self and world; forming a dialogue with images and landscapes; moving back and forth between image making and the dialogue with objects or landscapes; and reflecting and writing.

ii. **Magic** which refers to a special and exciting quality that makes something seem different from ordinary things; a parallel world next to the world of ordinary experience; the supramundane world of extraordinary experience where, like entering and exploring unique landscapes, languages and cultures, the domain of magic may be understood by observing young children at play or making art through deep attention and contemplation; this magical dimension involves a mindful shifting of one's perspective towards art. Indeed, art-making, in its highest form of expression, is a kind of magic. And in the magic of creation, the child becomes immersed in various art-making modalities.

iii. **Intuition** that explores the language of what one feels ... "a way of seeing through sensing", "an internal knowing of that which is invisible" (Woolery, 2006, p. 8); one draws on clues; one senses a pattern or underlying condition that enables one to imagine and then characterize reality; sensory understanding; intuitional and visceral ways of knowing; the bridge between explicit and tacit knowledge.

iv. **(Using) Imagination** becomes a modelling device through which we can test possibilities.

v. **Art-making** refers to the act of making art as a tactile event, a complete body experience where you feel the shapes inside your body and your body wields the tools that capture the shapes and, as a consequence, an artistic way of knowing can come about.

vi. **Language of pattern** addresses the dimensions of structure and pattern, focusing attention on shape, form, color, line, light and dark, value and pattern; shifting from macro to micro and foreground to background.

Methodologically, we purposively selected Art-Based Perceptual Ecology (ABPE) as our analytical framework and all six core dimensions (listed above) are clearly signposted throughout the chapter. ABPE gives a major role to key dimensions which are essential in understanding how to create the possibility spaces for STE(A)M education. The image that the teacher as artist, and the child as artist, produce can be seen as a symbol or transitional object that represents "the language of what one feels (*intuition*), with what one can touch (*direct experience*) with what one cannot see (*magic*) and *imagination* becomes a modelling device through which we can create possibility spaces" (Woolery, 2006, p. 13) (see Figure 11.1 where the ABPE methodology can be seen as framing the analysis and operationalising the process innovation in primary education).

Figure 11.1 UCPS, where releasing imagination and possibility thinking are central dimensions of the learning landscape.

Building on Hammersley, Gomm & Foster (2009), we utilise a comparative method of analytical deduction in relation to Woolery's dimensions or tools, looking, firstly, at what theory highlights about practice and, secondly, for similiarities and dissimiliarities in application. Comparative analytic analysis is open-ended; it goes beyond testing a hypothesis or identifying causal relationships. The purpose of the analysis of each practice was to elucidate the ways in which ABPE dimensions shed light on data that represents the language of what one feels (intuition), what one can touch (direct experience), what one cannot see (magic), what one intuitively knows and what is held as memory in human cognition, which the learning space enables through the experience of art-making.

With its mission – "releasing the imagination: celebrating the art of the possible" (which pays homage to Maxine Greene's social imagination work (Greene, 1998)) – the University of Cambridge Primary School (UCPS) (www.universityprimaryschool.org.uk), the UK's first University Training School for primary education, seemed to be the most obvious site for our multi-method project. UCPS features a circular design with a central courtyard which suggests the democratic notion of education. The school provides education for children aged four to eleven. It is an inclusive school.

Displaying works of art and artistic work and engaging artists in participatory approaches who work at the boundaries of formal and less formal learning, are common practices in this school, a site where possibility 'space', an enabling context for possibility thinking, is embedded in the school's design. As Maxine Greene (1978) argues, "landscapes of our interior worlds flow and merge into the landscapes of the exterior world" and "can stimulate much needed educational change" (p. 37).

Glass panels form a canopy around the circular design, featuring the work of Ruth Proctor[1]. Each glass panel is imprinted with an image of the sky taken from locations across the globe and stamped with time zone and geographical location. The art work itself is called 'We are all under the same sky'. Children at the school have engaged with these panels with both their teachers and artists. Dewey made clear the value of fostering and encouraging learners to value their own artistic cultures when he said "works of art are the most intimate and energetic means of aiding individuals to share in the art of living" (Dewey, 1934, p. 336).

11.1.2 Introducing the Installation 'With the Heart of a Child'

Commissioned by Anna Craft for the purpose of inspiring a vision of possible educational futures, 'With the heart of a child' is a sculpture installation described by its author, eco-artist/activist/sculptor Nicola Ravenscroft, as bringing together "all nations on Earth in the creation of a significant and innovative sculpture installation: it raises intellectual and emotional awareness of the urgent need to educate children and adults alike in the art of sustainable water management and conservation, and seeks peacefully to effect change, both to public policy and human behaviour"[2].

'With the heart of a child' embodies a powerful set of values and a message that lies at the heart of solving the pressing planetary problems shared by the global community. This powerful installation depicts six life-size bronze portraits of four-year old children (Figure 11.2) representing six continents and a rockhopper penguin-child representing Antarctica. It embodies two fundamentals of life on Earth – *water*, our most precious natural resource, and *children*, our source of continuity. The way in which the installation generates a new way of responding to these issues across the world remains under-theorised.

---

[1]Ruth Proctor is an artist. Her work responds to the spaces and time she lives and works within. See http://www.artspace.com/ruth-proctor

[2]See www.withtheheartofachild.com for more information about the artist, and the purpose and significance of this installation.

Figure 11.2 'With the heart of a child' installation.

The installation provided the basis from which to explore the significance and potential role of arts in developing new forms of STE(A)M education. From here, we articulated the research questions as follows:

1. What can the arts contribute to STE(A)M education in primary education?
2. How can the ABPE analytical framework provide a reference for understanding the enabling condition of 'possibility spaces' as a context for connecting STE(A)M in primary education?

The process of collecting, analysing and interpreting empirical data occurred between October 2016 and April 2017. Data were generated through collaborative discussions, classroom observations, and analysis of learning artefacts. This chapter is focused on the perspectives of teachers and children. These conversations fostered what Chappell and Craft (2011) described as a 'living dialogic space' where each of the researcher partners (co-authors of this paper) engaged in reflective enquiry practices to attain a deeper understanding of the meaning and point of the direct learning experience of art-making and thematic role of the arts in STE(A)M education. As Max van Manen (1990) explains, a theme is the experience of focus, of meaning, of point; it is a form of capturing the phenomenon one tries to understand: "Themes are the stars that make up the universes of meaning we live through" (p. 90).

As insider researchers, the translation of intuitively grasped meaning or tacit knowledge is not always possible through conventional and discursive renderings of language. Here we adopted an art-based research approach using ABPE methodology to interpret the role of art and art practice. The unit of analysis – the situated practices – and the data reduction involved the researchers narrowing in on three sets of data. Three data sets, covering both teachers' and children's perspectives, were collected and then subjected to extensive and focused analysis by all six researchers using the six core dimensions of the ABPE thematic analysis: **(i) direct experience, (ii) magic, (iii) intuition, (iv) imagination, (v) art-making** and **(vi) the language of pattern**.

These data sets were collected during the study over two school terms across 2016 and 2017. The six researchers, including an artist, several teachers, the Headteacher, a senior teaching assistant and university research partners were directly involved. Data were collected through (teachers', artists' and researchers') discussions, observations, teachers' self-reflection accounts, photos (taken by teachers), scrapbooks, workshops, artefacts, (children's own) audio-recordings and interviews. *Data sets 1 and 2* feature teachers' and children's exposure to, exploration of, and interaction with a piece of art – the sculptural installation 'With the heart of a child'. Both teachers and children were invited to work as artists, creating possibility spaces for figurative/abstract responses to the sculptures and for little-me making activity (van Boeckel, 2013). *Data set 3* features creation of possibility spaces, in a set of formal classroom practices, for the fusion of arts and mathematics, i.e. teaching and learning mathematics through seeing and hearing patterns, discovering one's body in a new way, thinking with one's hands, seeing music and hearing mathematics. The descriptive overview of the three data sets is presented in a tabular form below (Table 11.1). This is followed by a detailed description and analysis in narrative and visual form.

11.2 Analysis and Findings

As mentioned in the previous sections, art, art-making and art responding have been used to both collect data and interpret the role of art and art-making. The three data sets are presented, explored and analysed through the lens of the ABPE six core dimensions (direct experience, magic, intuition, imagination, art-making and the language of pattern) and their manifestation in different but interconnected and multi-layered practices.

Table 11.1 Overview of the three data sets

Purpose	Data Sets	Participants
	Data set 1:	
To observe the **teachers'** exposure to and interaction with a piece of art through art-making with an aim of inspiring didactic translation of their experiences and interactions with art into STEM lessons.	a) Non-participant unstructured **observations** of **12** primary teachers' initial exposure to the sculpture installation followed by teachers' **photos** and **self-reflection accounts** of their impressions, questions and internal dialogue. Purposeful engagement and interaction with the installation/the art. b) **Teachers' CPD workshop** with **12** primary teachers, using clay and water to produce their own pieces of art which they integrated with larger sculptures creating their own new installation, followed by teachers' **self-reflection accounts** of their experience. Teachers then engaged in possibility thinking about innovative lesson plans in STEM subjects.	a) 12 primary teachers b) 12 primary teachers
	Data set 2:	
To observe the **children's** exposure to and interaction with a piece of art through creative cross-curriculum activities with the aim of inspiring connections between their experiences and cross-disciplinary/ STEM learning.	a) Placement of the installation in a primary school – **teachers' observations** of children's reactions. b) **Art and Design project** inspired by the presence of the installation 'With the heart of a child' sculpture at a primary school – **observations** of the children's interaction with the installation and the art-making process of their own little-me 'sculptures' (**artefacts)** documented in **a scrapbook.** c) **Cross-curricular activities** involving Art (sketching), English (planning, writing, speaking, reading, drama), Computing (using digital media purposefully), Science (experiments with building and testing shelters made of different materials) – **teacher's observations, an interview** with the teacher and children's **artefacts** and **audio recordings.**	a) whole school b) 21 children and 1 teaching assistant, 1 artist (Nicky Smith), 1 governor/artist, 1 parent/artist (wider community) c) 25 children
	Data set 3:	
To observe and experience how STE(A)M **teachers** engage **children/learners** in creating possibility spaces through art-infused mathematics teaching/learning (seeing mathematics through art; hearing mathematics through music).	**Observation** of a Year 3 mathematics lesson, led by a specialist Finnish STE(A)M teacher, based on engaging with and creating arts and design-infused mathematics learning and a Year 2 mathematics lesson based on engaging with and creating an arts/music infused mathematics learning environment and interviews with the teachers.	49 learners 2 teachers 4 guest STE(A)M teachers

11.2.1 Teachers' Experience of and Interaction with the Installation (First Exposure to the Installation and Artist-Led Participatory Teacher Workshop)

The installation was placed at the UCPS and, as a non-gallery space, brought art closer to both teachers and children as well as providing inspiration for creating both physical and metaphorical 'possibility spaces' within a school setting. The first data set, realized as a participatory design workshop with teachers that was led by an artist, features the use of the sculpture installation 'With the heart of a child' to *inspire a co-created set of 'possibility spaces'* for STE(A)M education in the primary curriculum. Prior to the participatory workshop, the 12 UCPS teachers were invited to engage with the installation in a purely experiential way and, using their phone or an iPad, to take photographs/film/draw a particular part of the sculpture that attracted them. They were invited to note down any inner dialogue and intuitive responses.

Teachers had an opportunity to reflect and engage with the materiality, atmosphere and presence of the sculptures and, through this, reflect on the global position of children and water scarcity/sustainability. Teachers were able to touch and interact, on this occasion, something often prohibited in galleries or museums, but depth of engagement was not dependant only on the physical touch. The invitation was to respond to the sculptures, to be amongst them, to observe them, to notice their own inner dialogue (Figure 11.3).

Figure 11.3 Teachers' first exposure to, and interaction with, the installation.

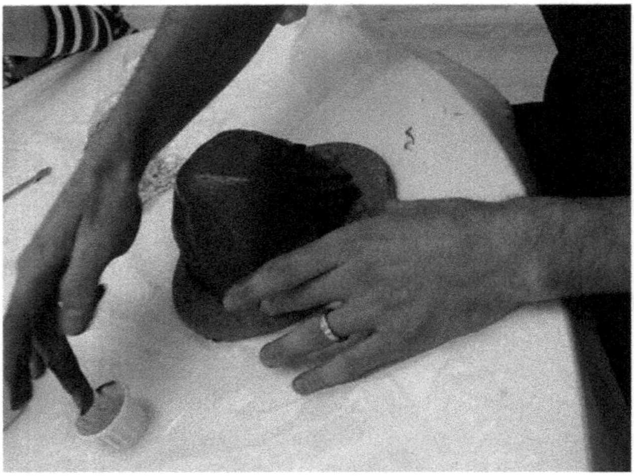

Figure 11.4 Teachers' art-making.

The second part of *Data set 1* encompassed art-making as the 12 UCPS teachers attended a 'clay workshop' where they engaged both with the installation, with the movement of the 'bronze children' around the room and with the process of art-making itself. Teachers were asked to use clay and water as essential components of the bronze sculptures (the process of making bronze sculptures often involves clay and water before metal is used), to make an image, an abstract shape, a person or an animal in relation to their initial impressions (Figure 11.4).

They were encouraged to allow the materials and their emotional responses to reveal the shapes and meanings of their creations. They placed their own sculptural responses among the 'bronze children', thus curating their own installation. Towards the end of the workshop, teachers were asked to explore how their experience of the installation and process of art-making could infuse, or be integrated in, already planned school curriculum themes. They were asked to consider the implications of using an arts-based approach with their children (e.g. in science – water and sustainability). Teachers discussed, devised and wrote ideas on paper and then placed them on the wall with links and lines connecting with school curriculum themes. After the workshop, teachers' reflective accounts of their experience were collected.

Table 11.2 First data set's presentation and analysis

ANALYSIS OF DATA SET 1 (Teachers' Initial Exposure to a Piece of Art and an Artist-Led CPD Teachers' Workshop)	
Illustrative Examples	ABPE Core Dimensions and Specific Findings
 **Figure 11.5** One of the teachers focused on facial features. *Quite haunting. The lack of expression and blurred facial features make* ***me*** *feel a little* ***un-nerved*** (Teacher A)	Teachers' **direct experience (i),** exposure to and observation of the sculptures evoked *emotional states:* • un-nerved • hope/hopeful • confident • determination • excited • compassion • joy • powerful • submissive • scared • dried out • burnt • sadder • despair In the written reflections on their **(i) direct experience** of the installation, teachers linguistically illustrated *connection* between the sculptures and their own selves by alternating between writing about the sculptures (e.g. the children, they, them) and about the impact the sculptures had on teachers' own emotional experience (e.g. makes *me* feel, *I* felt excited, a very positive impact on *me, I* feel instant joy). Teachers described both the sculptures' and their own *emotional states.* The combination of *experiencing* emotional states and the relational dynamics between themselves and sculptures led to creation of *experiential possibility spaces* for *making meaning* out of connections between the 'bronze children' (others/world) and self.
Hope*, feeling* ***hopeful****, looking to the future, places to go* ***Confident*** *children ready to take on/explore the world* ***Our*** *children can achieve anything* ***Determination*** *The children had a very positive impact* ***on me****.* ***I*** *felt* ***excited*** *by them and wanted to know more.* (Teacher G)	
Upon greeting the children for the first time ***my*** *immediate* ***emotion*** *was that of* ***compassion*** (Teacher C)	
At first, ***I*** *feel instant* ***joy*** *on viewing the statues. The children look* ***powerful*** *. . . the more* ***I*** *look at them, the more* ***I*** *think about the possibility of them being* ***submissive*** *and* ***scared****. They look* ***dried out*** *and* ***burnt****.* ***I*** *am feeling* ***sadder*** *and* ***sadder*** *. . . It looks more like them looking up in* ***despair*** *now.* (Teacher F)	

(*Continued*)

Table 11.2 Continued

<table>
<tr><th colspan="2">ANALYSIS OF DATA SET 1
(Teachers' Initial Exposure to a Piece of Art and an Artist-Led CPD Teachers' Workshop)</th></tr>
<tr><th>Illustrative Examples</th><th>ABPE Core Dimensions and Specific Findings</th></tr>
<tr><td>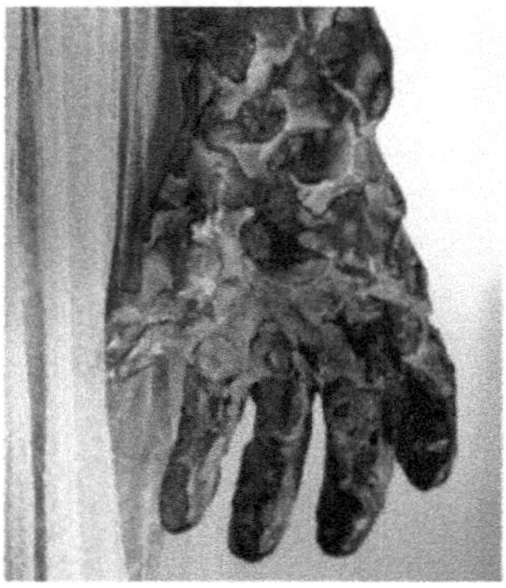
Figure 11.6 Teacher's focus on 'bronze child's' hands.

Our tiny hands must carry the same hope.
We crumble under the same need.
Our sleep is haunted by the same dreams.
Our hands are all the same. (Teacher B)</td><td rowspan="3">While taking photos of parts of the sculptures, teachers focused on their body parts (e.g. faces, lips, hands) and were interpreting them through seeing, hearing and feeling what was not explicitly there:
• "Our sleep is haunted by the same dreams"
• "Our tiny hands must carry the same hope"
• ". . . when children have no voices?"

Teachers also used 'we' and 'our' to refer to both the sculptures ('bronze' children) and themselves clearly indicating unity and care and concern:
• "Are you ok? Where are your parents/carers? Do you need a hug?"

Using (iv) imagination, (iii) intuition and allowing themselves to feel a 'magical moment' in a (ii) magical space, led teachers to (iii) intuitively experience what was not explicitly there thus opening intuitive possibility spaces for the feeling of unity, concern and care and bringing otherness closer to self.</td></tr>
<tr><td>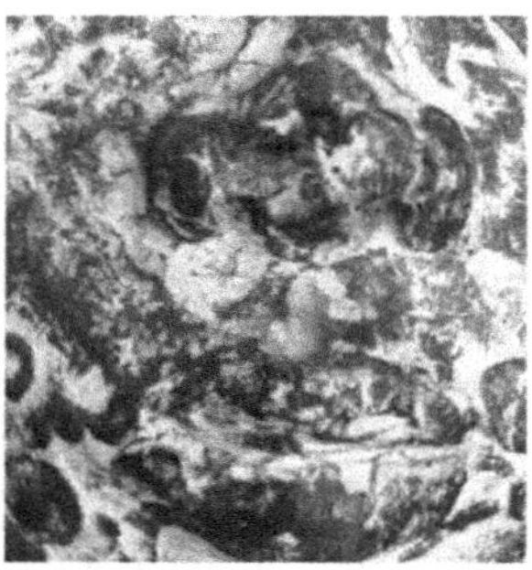
Figure 11.7 Teacher's focus on 'bronze children's' lips.

How can we release voices and open eyes to possibilities when children have no voice? (Teacher C)</td></tr>
<tr><td>Are you ok? Where are your parents/carers? Do you need a hug? (Teacher A)</td></tr>
</table>

Figure 11.8 Teachers working with clay and adding to the installation.

... both working with the clay and adding to the installation made me more excited about the sculptures. I ***felt more a part of them*** *and my understanding and* ***relationship*** *with the sculptures changed. The next day, I brought my class to feel the sculptures as we were exploring bronze as a material in our history studies.* (Teacher B) – reflective account)

Using and relying on the senses, is not something I like. I am not comfortable to enter into ***oneself****.* (Teacher F)

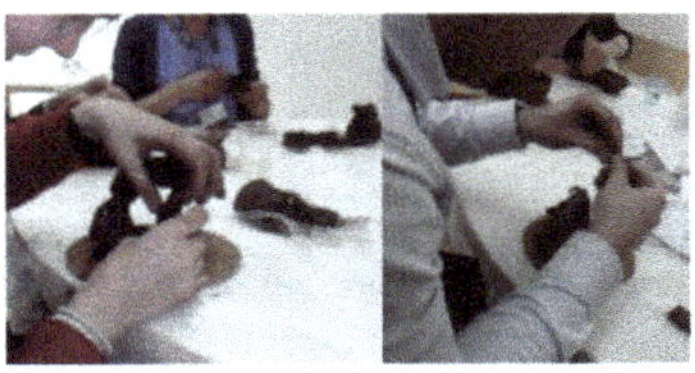

Figure 11.9 Teachers exploring materials used for art-making.

Working with this clay has deepened my understanding of the art work. In particular, through working with the clay and handling this material I have reconsidered the ***relationship*** *between clay, water and sustainability.* (Teacher A – reflective account)

Figure 11.10 Teachers interacting with the installation.

Teachers experienced an artist-led Continuing Professional Development '(CPD) programme for teachers through **(v) art-making** as a tactile, open-ended, sense-based process. While using their hands, teachers fostered the appearance of *embodying possibility spaces* for *relational inter-connectedness* (e.g. with 'oneself', with the children/others, with 'the sculptures'/a piece of art, with and across disciplines and subjects).

The **(v) Art-making process** inspired teachers to 'blur' dualisms between others and self, between the art-making process and a piece of art, between different groups of disciplines and subjects (e.g. using the installation for history, science, mathematics lessons) and build 'connections' across subjects.

(*Continued*)

Table 11.2 Continued

ANALYSIS OF DATA SET 1 (Teachers' Initial Exposure to a Piece of Art and an Artist-Led CPD Teachers' Workshop)	
Illustrative Examples	ABPE Core Dimensions and Specific Findings
. . . I hope to use the ideas stimulated from this session within my planning. Specifically, I will consider the ***connections*** *that can be made across subjects more carefully and creatively, as a result of . . . interacting with the artwork.* (Teacher F – reflective account)	
The workshop has made me consider how we could plan a series of 'what if' sessions with the children working in small groups. (Teacher G – reflective account)	
. . . We could also use this as an opportunity for the children to think in scientific ways to solve the challenge of not having clean water. It would also be possible to discuss distance, measuring and capacity with the children through this activity. (Teacher B – reflective account)	
This could lead to map making, story writing, film making, stage directing, dreaming and imaging. A soundscape could supplement all of these ideas. (Teacher C – reflective account)	
Clay *Move* *Mould* *Drip* *Drop* *Life* *Flow* *Forever* *Never* (Teacher D – reflective account)	Teachers' exposure and interaction with art provided *exploratory possibility spaces* for the appearance of the **(vi) language of pattern**: *internal and external patterns* in the form of, on one hand *expressing internal dialogues, thoughts and feelings* and on the other hand, discovering *external patterns and procedures of art-making* thus exploring both foreground and background patterns, visible and invisible patterns, physical and abstract patterns.
. . . Splashing colour on *On* *On* *Tired faces* *On* *Breathing in silently* *Deeply* *Reflecting on what why* *Why here in a school* *Why* *Silent teachers . . .* (Artist's reflective account while observing teachers' initial exposure to the installation.)	

In summarising teachers' direct experience of and interaction with the sculpture installation, as with findings from the teachers' clay workshop and, in this case, the role of the installation in creating possibility spaces, we see evidence that:

a. Through **(i) direct experience** of exposure to and interacting with a piece of art, teachers engaged in a curious, emotional and reflexive exploration of *experiential possibility spaces* for *making meaning out of sensation* through attempting to understand the *connection between the 'bronze children' (others/world) and self* (e.g. through evoking different emotional states from despair to hope). Teachers used descriptive narrative to express emotionally their **(i) direct experience** of the installation, and alternated between talking about themselves and the children (by using 'I'/'me' and 'children'/'them'/'they') thus demonstrating relational dynamics between 'others and self'.
b. While experiencing and interacting with the installation, teachers used **(iv) imagination, (iii) intuition** and found themselves in a **(ii) magical non-gallery space** where they **(iii) intuitively** used either 'we' sentences (expressing unity of themselves and the sculptures/'bronze' children) or direct speech sentences (addressing the sculptures/'bronze children directly with 'you-sentences' when expressing concern and care for them). Direct speech was used when **(iv) imagining** and **(iii) intuitively** seeing and sensing what is not explicitly there (e.g. voices or lack of them, sleep or lack of it). Thus, *intuitive possibility spaces* for the *feeling of unity, concern and care* and bringing *otherness* closer to *self* seemed to gain significance. The sense of **(ii) magic**, a special and exciting quality that makes something seem different from ordinary things, that arose in the **(iv) imaginative** and emotive responses to the sculpture, was a sense in the imaginative possibilities inherent in teaching and learning; that there was wider significance in the work of teachers, beyond practical matters or accountability frameworks, but something more akin to the social **(iv) imagination** that Maxine Greene advocates (1998). There was a sense of **(ii) the magic** in the *intuitive possibility spaces*.
c. While changing roles from being observers of art to active engagement with art through **(v) art-making**, teachers focused on relationship and connections with oneself, with each other, with the process, with the art piece and with different subjects through contemplating how to 'translate' their experience into a cross-curriculum/trans-disciplinary learning space for children. Thus, the *embodiment of possibility spaces*

for *relational inter-connectedness* (e.g. with oneself, with others, with a piece of art, with and across disciplines and subjects), was fostered.

d. While exploring and making art, teachers discovered *exploratory possibility spaces* for expressing and discovering the *internal and external* **(vi) language of pattern** through expressing internal dialogues, thoughts and feelings and through exploring external patterns and procedures of art-making.

The teachers' reactions to, exposure to, and interaction with art were aligned directly with the affective, sensuous and embodied dimensions of learning where physicality and associated feelings shape learning (Claxton, Lucas & Webster, 2010; Fuchs & Koch, 2014; Kontra et al., 2015; Stolz, 2015). They experienced art directly and actively. The teachers experienced the process of creating possibility spaces, with the point of departure being a response to the art installation, and in this process, discovering how sensory perceptions can generate STE(A)M practices.

11.2.2 Children's Exposure to and Interaction with the Installation

The second data set explores the dynamic relational and living space between the installation 'With the heart of a child' and the children at UCPS.

The initial location of the installation in the school hall/gym (the bronze statues arranged in a line looking down from their plinths) was designed to provoke an open and child-led response. Initially, the learning was about asking questions – any question, all questions.

i. Direct experience: Similarly to teachers, children's first exposures to the installation opened *exploratory possibility spaces* for evoking *emotional states* as illustrated below by teaching assistant's observation:

> *Some of the children found them disconcerting: 'they are dead', 'they look sad', 'they look poor'. 'Why are some people poor?' The statues seemed to have an effect on the children because parents came to talk with the Headteacher saying that we needed to do more work with the children so that they didn't feel worried about the statues ...* (Senior teacher assistant's observation)

The children viewed the statues in many different arrangements/positions. The installation was later moved to a smaller space to change the dynamic and qualitative features; it was moved to a special room named the 'the Nicola

Ravenscroft room', which had a sign on the door reading 'Nicola Ravenscroft and The University of Cambridge invite you to enter peacefully with an open heart and mind'. Children were encouraged to open their hearts and minds, thus reinforcing, this time, positive emotional states with which to approach **(i) direct experience** of the installation. Children's emotional states moved from being scared and worried (*'they look poor')* for the 'bronze children' to entering the installation room with open hearts and minds. Thus, *making meaning* out of connections between the 'bronze children' (others/world) and self contributed to the creation of *exploratory possibility spaces* through the **(i) direct experience** of art.

vi. Language of pattern: The room/place contained all the 'bronze children' arranged in a different constellation – this time in an arc and at ground level. Different shades of blue material to create the impression of a vast ocean were used. As well as this, books about water, water-related pictures and a water music (CD playing in the background) were added (see Figure 11.11). The UCPS team clearly set out on a journey of arts-based perceptual ecology by bringing an awareness of *patterns* in the 'landscape' and changes in locations, positions and effect of art and how all these yield the language of place. Children experienced *exploratory external patterns* of different physical arrangement, location and positions of the installation and reacted differently to the installation based on its location and arrangement. The new curation was a provocation to teachers and children to ask more questions. The more intimate arrangement provoked different responses to the installation. New possibilities in new possibility spaces arose.

Figure 11.11 The installation room's floor turned into 'water'.

Figure 11.12 Final 'sculptures' made by Art and Design Club members.

v. Art-making: The Art and Design Club/Project was inspired by the 'visit' to the 'With the heart of a child' installation and, through self-selection, 21 children from Years 1–3 enrolled in the club. Children spent a few weeks working on creating their own 'little-me' sculptures, using foil and Modroc, with an overarching theme that included ideas and plans about future and 'All about me'.

All the activities, ranging from creating artefacts to writing about themselves, were recorded in the form of a scrapbook. Over the period of a few weeks, children learned how to make a 'sculpture/little-me' of themselves in relation to their hoped for/perceived future by first using their own bodies to explore different stances, then manoeuvring wire and foil to create sculptures and finally completing Modroc 'figures' (Figure 11.12). The use of their bodies prior to art-making enabled children to explore art through *embodiment* of their future creation as documented by the Senior Teacher Assistant (TA) below:

> *Firstly, they were encouraged to think about their own bodies and how they could stand in different poses. The children stood up and practised standing in different stances. Next, they were given two pieces of foil to manipulate into a foil figure. All of the children set to work to create some amazing foil figures, all of the poses were so different.* (Senior TA's observations)

Thus, again, similarly to the teachers' experiences, through a tactile, hands-on process, the children experienced embodying *possibility spaces* for art-making. The embodied responses were in connection to the *creation* of the world of the future rather than just *inheriting it.* It was in an empowering future of possibilities rather than in an apocalyptic vision of uncontrollable certainties that the children engaged in positive explorations of their futures.

> *I am blown away by the level of concentration, thoughtfulness and pride these young children demonstrated during the project* (Senior TA's reflective account)

Using (iv) imagination, (iii) intuition and (ii) magic: Children also explored many art-based and art-inspired activities while visiting the installation room. A spectrum of teacher-designed activities took place, ranging from observation, sketching the installation, imagining the bronze children were a travelling circus troupe that ended up on an island, taking the identity of one of the bronze children and writing a story (Figure 11.13) about what happened to them from a first-person perspective.

When children turned their stories into 'first-person' stories and narrated them as if they were the 'bronze children' their teacher noted that they gave 'bronze children' voices; this builds nicely on what teachers asked in their very first exposure to the installation (see Table 11.2).

> *How can we release voices and open eyes to possibilities when children have no voice?* (Teacher C)

In order to take on an identity of one of the 'bronze children', children used **(iv) imagination** and described **(iii) intuitively** what might have happened to the 'circus troupe' that ended up on an island. Children saw what was not there and sensed what might have been there. As they became absorbed by the story that something happened to the circus troupe, they wanted to know what happened next, and this was then handed over to them. They were imagining, seeing what is invisible and hearing what is inaudible: who the children are, where they went, what happened to them – thus answering the questions that both teachers and children in their first exposure to the installation asked. Art-making (of sculptures or stories) thus became a means to reveal *embodied*

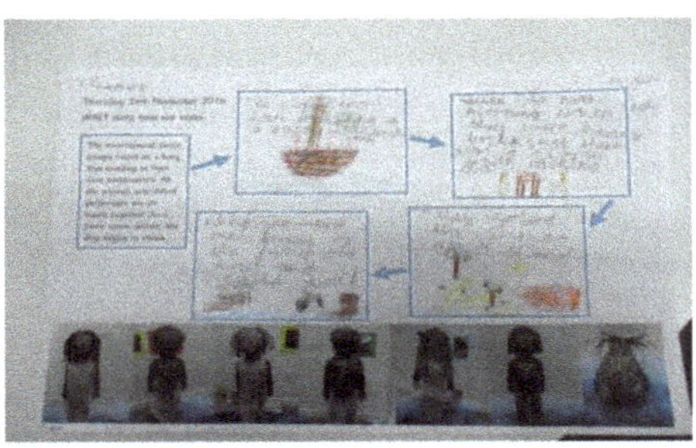

Figure 11.13 Story mapping.

intuitive knowing. The temptation to answer their questions was avoided by teaching staff, as one of them recalls:

> *Being confused was an important part of the process and was embraced. Staff resisted answering questions but engaged children in finding their own solutions ... or sometimes in recognising that problems could not be answered ... but required more questions ...* (Senior TA's reflective account)

Thus, teachers managed to transfer *agency* to children, who, in groups, were making decisions on their own on which direction to go with their stories and what events they wanted to include. Again, similarly to teachers, children experienced *intuitive possibility spaces* for the *feeling of empathy* (while imagining being 'bronze children' and building Lego-bricked boats and shelters for them) and bringing *otherness* closer to *self,* and adding the notion of *agency* to their use of **(iv) imagination, (iii) intuition** and the **(ii) magic spaces** they created.

11.2.3 Trans-Disciplinary Learning Spaces

Once children's stories were finished, they were peer assessed and, later, children used modern technology/applications to record sounds of water running, a door slamming, etc., to accompany audio recordings of their stories, which they subsequently stored in their folders in Dropbox (Figure 11.14).

During the phase of children's exposure to and interaction with the installation, teachers, researchers, teacher assistants, governors and parent volunteers were involved in the project. A Senior TA summarized the phenomenon:

> *I loved this element – not only had we had a governor and the university but also a parent involved – the project had a real community feel.* (Senior TA's reflective account)

The final sculptures/mini-me's children made were displayed in the art room window. The collection was named 'From the heart of a child'.

Adults involved in the project shared their surprise at how well children worked on diverse trans-disciplinary activities as well as on how well they incorporated the themes of water and sustainable futures into their sculptures and stories.

> *It was amazing that independently the children's plans for their sculptures contained elements of water that linked so well with the installation.*

> *Linking the statues with the environment was a direct link between art and geography, and matters of climate change. Children were involved in a number of activities that required them to continue asking questions.*
>
> (Senior TA's reflective account)

Children's trans-disciplinary (STE(A)M) learning was summarized by teachers in the following subject-related categorization, which represents the far-reaching nature of arts-infused practices on cross-curriculum learning.

Figure 11.14 Children using technology to audio record sound effects for their stories.

Table 11.3 Overview of children's learning experiences during many arts-infused and arts-based practices/cross-curriculum activities inspired by the presence of the installation

Subject/Area	What Have Children Learned?
Art	• observational sketching • personal response to a piece of art
Engineering/Design	• story mapping/planning structures • writing an account of modelling • seeing designs in nature • mapping the body/connecting to design
Technology/Computing	• using and manipulating digital media purposefully • using software for audio recording • storing and organising digital data
Science	• making simple observations • making predictions • building physical structures (with Lego) • seeing the science in nature
Mathematics	• taking mathematical roles • storying problem solving using mathematics outdoors • making sense of numbers (performing patterns, shapes)

Children's direct experience of, and interaction with, the sculpture yielded interesting findings similar to those of the teachers, indicating that exposure to the installation initiated the appearance of similar *possibility spaces* across the school setting. From the analysis of the children's exposure to the installation, we see evidence that the following *possibility spaces* were opened:

a. Through **(i) direct experience** of the installation children found *exploratory possibility spaces* for evoking *emotional states* (from fear and concern to openness of heart and mind) that brought about the understanding of connections between the 'bronze children' (others/world) and self.

b. Using **(iv) imagination, (iii) intuition** and creating **(ii) magical spaces** helped children experience *intuitive possibility spaces* for the *feeling of empathy and agency* which brought *otherness* closer to *self.*

c. Through the tactile, hands-on process of **(v) art-making** children experienced *embodying possibility spaces* for art-making in connection to *creating* the world of the future rather than just *inheriting it.*

d. Through being exposed to the installation in different locations, positions and arrangements children experienced first-hand *exploratory possibility spaces* for the appearance of the **(vi) language of pattern,** and learned the importance of landscape/place patterns.

What children learned during their exposure to, and interaction with, the installation reinforced what was mentioned at the beginning of the chapter – that the arts can enhance high performance teamwork, improve observational skills and adaptability. Children (similarly to the teachers) moved from the *direct experience of the emotional state of fear* and thinking that the 'bronze children' are dead to *caring* about them and building them boats and shelters, thus clearly supporting the belief that the arts are social, inclusive, humanising and thereby significant for human development in society (Belfiore & Bennett, 2007; Canatella, 2015). There seemed a movement from inertia to social activism: the children moved beyond their fear of the unvoiced 'bronze children' to providing their own voices through the 'bronze children' and for the 'bronze children'; from uncertainty in experiencing the static sculpture to actively creating new imagined journeys for the sculpture.

Through the changes of the locations and positions of the installation, children experienced it differently in a 'special room' where one entered with an open heart and mind. Arts-based perceptual ecology again proves to be valuable in understanding how places, environments and positions can be important in discovering patterns of interactions and relations between people and their environment.

11.2.4 Teacher-Led Didactic Formal Classroom Session

Part 1: 'Seeing' mathematics through constructing structures

The third set of practices is divided into two parts. The first classroom session features the use of arts-infused practice to teach mathematics through constructing/building models for a Year 3 class (children 7 to 8 years old) and the second classroom session teaches Year 2 mathematics through music. In the first session, led by a group of Finnish specialist STE(A)M teachers, creative meaningful learning spaces use art-making not as an object but as a playful event to teach pentagons, hexagons and patterns and shapes one can create with them. The lesson's goal was the experience-oriented discovery of a truncated icosahedron (which is the structure of a football or the Fullerene molecule), learning about connections between 2 and 3-dimensional patterns and structural manipulations of pentagons and hexagons.

(i) Direct experience and (v) art-making: Children were exposed both to real balls (football) and photos as well as real 'building materials' and model samples and photos of them (Figure 11.15). This prompted children to construct their own 'football'. The teacher introduced the topic through questions about what children thought they could make out of the presented materials and kept the whole lesson focused on creating, building and discovering patterns, which led to children creating 'houses', sunrays, etc. Through inviting children to first explore available materials, then to see different models made of the same materials and finally to compare the physical ball with models, the teacher provided conditions for children's **(i) direct experience** of both materials and models. In a similar way to that of the children who experienced *experiential possibility spaces* for *making meaning* out of connections between the installation and self, the Year 3 children experienced *experiential possibility spaces* for *making meaning* out of connections between the models and self.

(v) The art-making process was present throughout the lesson and mathematics learning happened alongside creating possibility spaces for arts-infused practice. Children stayed excited throughout the lesson and were learning mathematics through their 'hands', while constructing/building their own 'football' models in an open-ended task. *Embodying possibility spaces* for art-making connected to *constructing* mathematics rather than just *learning it* involved some risk, as one of the teachers noted:

> *To me, it was left open. I had no expectations of what they needed to meet. It was not said. It was not prescribed.*

We shared our worlds together for a short time. Something was and is always at stake for the participants, the teachers, the children, we all took a risk. The risk is that we all come upon a certain understanding that was not given on forehand. It's not about understanding of information being processed. It's, in essence, educating new insights. (STE(A)M Teacher)

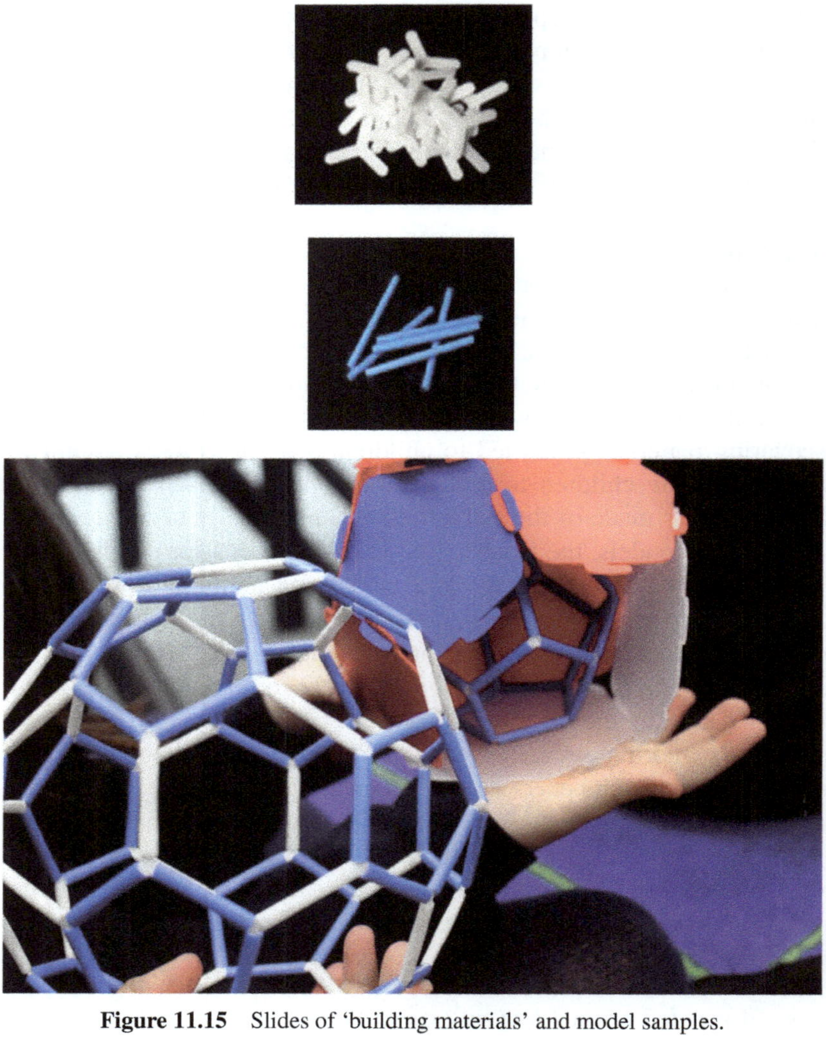

Figure 11.15 Slides of 'building materials' and model samples.

Using (iv) imagination, (iii) intuition, (ii) magic and **(vi) the language of pattern:** The teacher's invitation to children to transform into 'inventors' and to think differently opened the *intuitive possibility space* for **(iv) imagination, (iii) intuition** and **(ii) magic** (i.e. seeing, hearing and feeling what is not there). In no time, children took on the identities of inventors and came up with ideas as the illustrative examples below show:

> *I'm making a snowflake.*
>
> *I'm making a house.*
>
> *I'm following my own way. I have made two balls. One is inside the other.*
>
> *I have made planet earth and an earth station.*
>
> *I made a doghouse.*
>
> *I made something for people who are blind ... it works well ... it's all about feeling it.*
>
> *I made a bell. It was really hard for me to kind of get what was going on. I concentrated on my hands. I closed my eyes. Ideas came.*
>
> (Researcher's notes)

Children worked in pairs, studied thoroughly the **(vi) language of pattern** of real balls in the classroom, and each made a half of the ball. They supported each other by checking whether *patterns* in their halves were equal and could be used to form a whole ball with pentagons and hexagons. Direct experience of *pattern detection and pattern recognition* was enhanced with cooperation and team work similar to a group work of children creating a story about what happened to 'the circus troupe/bronze children'. In the children's words, they 'followed their own way', 'closed their eyes' and 'it worked well' because 'ideas came' and it was 'all about feeling'. Once again, perception was the key to *reading patterns* whether they are mathematical (arrangement of pentagons and hexagons in a truncated icosahedron) or emotional (shifts from one emotional state (e.g. despair) to another (e.g. hope)).

Compared to other data sets, which featured direct experience of art out of formal classroom settings (school hall, special room for the installation, art room), the mathematics lesson was delivered in the classroom. Before the closing discussion the children released all their energies piled up during the highly-concentrated building process whilst also focusing on the realization of their ideas: in the school's sports hall, free ball play allowed the children to enjoy their constructions. According to ABPE, each place has its own patterns. The teacher's invitation to children to use their **(iv) imagination** and **(iii) intuition** and explore how to think and build as inventors seemed to have removed any obstacles to creativity. The children's agency and propensity to seeing, hearing and feeling what is not there were opened up through the invitation.

Part 2: Hearing mathematics through music

The 'mathematics through music' lesson featured a Year 2 class whose teacher created meaningful *possibility spaces* throughout a day of 'performative musico-mathematic knowing', bringing diverse configurations of mathematics and music into meaningful relationships through STE(A)M education.

The mathematics teacher aimed to help children develop a 'mathematical lens' through which to interpret their experiences of the world; to consider mathematics as a subject imbued with patterns, and help children to make connections between maths, music and art. Lesson sequences built on these different contexts to bring understanding and insight into structures of counting, place value and multiplication. The lesson was part of a whole-day, whole-school event where different teachers engaged in a 'Musical Mathematics Day' and performed, sang, danced, heard and composed mathematics.

During the observed lesson children engaged in different activities ranging from learning to conduct and practising multiplication through clapping to composing through the use of three-dimensional mathematics (e.g. Numicon) and digital media (online applications).

(i) Direct experience: The mathematics lesson started with learning to conduct through listening to classical music rather than with an overt focus on mathematics. Children focused on conducting the pulse of the music (Figure 11.16), which gave them a way to actively participate in joining in with the music and making some implicit and explicit links to the relationships between music and number. The teacher reported that children were captivated by the music and enjoyed conducting.

After a musician (flautist) visited the class and introduced the children to the metronome, they played 'clapping times' in pair thus practicing multiplication through coordinating. The role of body (i.e. embodying mathematical

Figure 11.16 Children conducting the pulse of music.

concepts (e.g. multiplication)) required a cooperative learning focus as both players had to focus on their own numbers at the same time as reciprocal partner work.

Some children shared their experiences:

> *I hear intervals*
>
> *I see the intervals in sound now*
>
> *There is a different kind of surface to this sound when it's a colour or a line you've drawn ...*
>
> (Researcher's notes)

(i) Direct experience of listening to both the recorded and live music performances while learning mathematics created *emotional states* (enjoyment) that lead to the creation of *experiential possibility spaces* for *making meaning* out of connections between the musical patterns and numbers.

Art-making, imagination, intuition, magic and the language of pattern
Children were also invited to **(v) make art**, i.e. to compose by using the Cuisenaire rods and Numicon shapes (i.e. by using their hands and manipulating plastic pieces). They first explored making a simple rhythm using different number pieces to represent the number of notes they would play. Each child chose a part and **(iv) imagined** what kind of piece they would like to create. Later in the day this was extended for them to try overlapping two

parts which provided an opportunity for them to see number relationships and **(vi) patterns**. Conceptual variation of differing representations appears to help strengthen the understanding of concepts (Gu et al., 2004). Pattern recognition in art, landscape, environment and mathematics may be one of the commonalities across disciplines and as such is easy to transfer to all STE(A)M areas.

The **(v) art-making** process of making music by using plastic 3D pieces combined with the **(iii) intuitive** choice of **(vi) pattern** contributed to the *embodying, exploring and intuitive possibility spaces* for a holistic experience of *hearing mathematics* rather than just *learning it.*

The lesson finished with using a simple flash music animation online application that children explored and then, through trial and improvement, created a piece (Figure 11.17). The task involved problem-solving, reasoning, collaborative learning and **(vi) pattern** recognition.

According to the ABPE framework, in both lessons teachers and children were *co-constructing* and *co-composing possibilities* for STE(A)M education through:

a. *experiential possibility spaces* for **(i) direct experience** of models and music composing/constructing in mathematics
b. *intuitive possibility spaces* for the use of **(iv) imagination, (iii) intuition and (ii) magic** in mathematics
c. *embodying possibility spaces* for **(v) art-making** of composing/constructing in mathematics
d. *exploratory possibility spaces* for the **(vi) language of pattern** in relation to its detection and recognition in mathematics

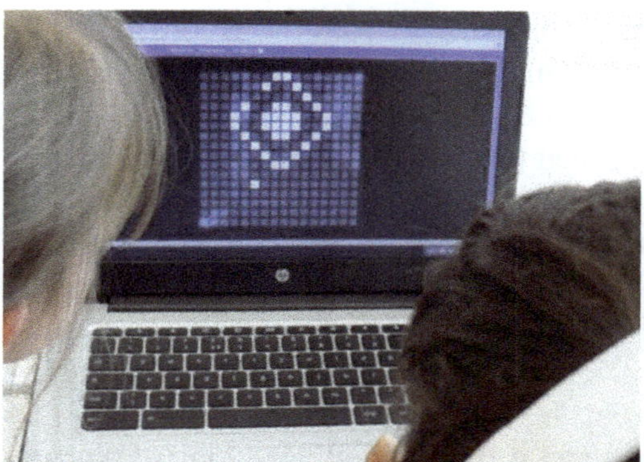

Figure 11.17 Children composing with the help of an online application.

11.3 Concluding Discussion

Along with offering leading-edge practices using STE(A)M Education (in Science, Technology, Engineering, Arts and Mathematics), this small-scale study promised several things: Firstly, we aimed to arrive at new understandings about what characterizes *'possibility spaces'* within the learning engagement of young children and how such spaces enable innovative practices. Secondly, we asked what the arts can contribute to STE(A)M education in primary education. Thirdly, we reviewed how the dimensions of ABPE methodology (involving direct experience, magic, intuition, imagination, art-making and the language of pattern), can be understood in several ways. Finally, we investigated the role that using a sculpture installation as a stimulus played in implementing the ABPE methodology.

In this chapter, we have further *theorised 'possibility spaces'* as the unique enabling spaces in which innovative learning engagements emerge. What these spaces and related practices highlight, in particular, is how a multiplicity of voices in a school context, evoked through the encounter of direct experience and art-making, can spur the imagination, ignite a sense of intuition and manifest as magic, acting upon, and situated in, the sensation of learning engagement.

So, what have we learned? We have learned that the experience of art-making provides a doorway which, once passed through, can help make new and valuable innovative links and meaning in Science, Technology, Engineering and Mathematics (STEM) education for both teachers and children. John Dewey held that an art form "carries the experience, not as vehicles carry goods but as a mother carries a baby when the baby is part of her own organism" (Dewey, cited in Woolery, 2006, p. 33). We have learned that in the intuitive process, the child finds a way of seeing through sensing, as an internal knowing which is art-making. We have learned that intuition stimulates imagination, which acts as an organizing process that creates representations of our learning experiences. Thus, imagination becomes a modelling device through which we can test possibilities and co-create enabling possibility spaces. We have learned that art-making is not simply a tool, it is a process which is a tactile event, a complete body experience where children feel the shapes inside their body and their body wields the tools that capture the shapes. We learned that if teachers engage in/with/through the co-creation of possibility spaces as sites of transdisciplinary learning, together with children, they give way to and in to 'possibility thinking' and teaching differently.

The central contribution of the arts-based perceptual ecology framework (involving direct experience, magic, intuition, imagination, art-making and the language of pattern) is twofold. Firstly, it offers a way to avoid dualisms like mind/body, self/other, thought/feeling. Secondly, the APBE methodology enables us to contain the great complexities and contradictions of learning engagement for STE(A)M education without diminishing them. We have shown why ABPE can be approached as a 'living system' (away from mechanistic approaches to primary education).

What can we take away from these findings which have significant implications for developing pedagogies of possibilities in primary education practice? We have illustrated how art-making is a unique means to acquire knowledge and how an artistic way of knowing can come about in primary education as part of STEM. If a sculpture installation can embody an idea then we need to accept that art-making involves perception, emotion and action – that is, the entire process of life. The artistic process, as with the 'little me' clay making that evoked responses to the sculpture installation by both teachers and children, becomes transformative. This shows the extent to which participating in art-making can make a profound impression on teachers, the children and, potentially, their communities. The sculpture installation invited participants into 'possibility spaces' where they came to see 'learning others' as learning selves and the significance of *how art creates a world*. The implication for creating sustainable futures is that art-making is a unique means to acquire and absorb new knowledge; it allows us to *know* ecological elements of the landscape in which we exist, and from which we cannot separate ourselves.

What we have come to understand about the complexity and materiality of teachers' work is captured in the concept of arts-based perceptual ecology, as pedagogies of possibility which facilitate innovation and change in co-creating transdisciplinary learning and possibility spaces for STE(A)M education. It is the dynamic *possibility spaces* which enable teachers to imagine possibilities for STE(A)M-oriented possibility thinking that provides new directions for pedagogies of possibility.

Why we consider that the study and the practice of arts-based perceptual ecology is important in our day and age is that it facilitates innovation and change, enabling possibility spaces for diverse configurations of STE(A)M education. We can conclude that STE(A)M education holds within it a new set of *possibility spaces* and opportunities for authentic, hands-on, interactive learning, allowing children, their teachers and school communities to form the vanguard of transformation now needed to co-create and realize their collective dreams for our sustained creative futures.

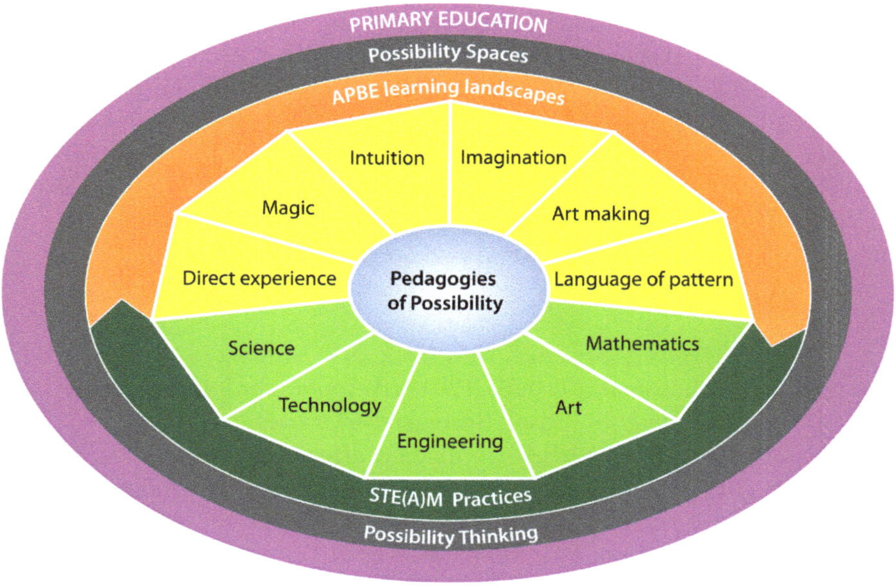

Figure 11.18 Towards a framework for innovation and change in primary education.

Acknowledgements

The research presented in this article was supported by seeding funds from the revenue of the 2nd International BiBACC conference. The authors wish to thank the participating UCPS community and Nicola Ravenscroft whose installation found its home at this site for 2 terms. We dedicate this project and paper to Anna Craft who inspired and enriched our lives.

References

Belfiore, E., and Bennett, O. (2007). Rethinking the social impacts of the arts. *Int. J. Cult. Policy* 13, 135–151.

Burnard, P. Craft, A., Cremin, T., et al. (2006). Documenting 'possibility thinking': a journey of collaborative enquiry. *Int. J. Early Years Educ.* 14, 243–263.

Canatella, H. (2015). *Why We Need Arts Education. Revealing the Common Good. Making Theory and Practice Work Better*. Rotterdam: Sense Publisher.

Chappell, K., and Craft, A. (2011). Creative learning conversations: producing living dialogic spaces. *Educ. Res.* 53, 363–385.

Chung, S. K. (2007). Art education technology: digital storytelling. *Art Educ.* 60, 17–22.

Claxton, G., Lucas, B., and Webster, R. (2010). *Bodies of Knowledge: How Learning Sciences could Transform Practical and Vocational Education.* London: University of Winchester.

Craft, A. (2015). *Creativity, Education and Society: Writings of Anna Craft Selected by Kerry Chappell, Teresa Cremin and Bob Jeffrey*. London: IOE Press.

Dewey, J. (1934). *Art as Experience*. New York, NY: Minton.

Fenyvesi, K. and Lähdesmäki, T. (eds). (2017) *Aesthetics of Interdisciplinarity: Art and Mathematics.* Basel: Springer-Birkhäuser.

Fuchs, T., and Koch, S. C. (2014). Embodied affectivity: on moving and being moved. *Front. Psychol.* 5:508. doi: 10.3389/fpsyg.2014.00508

Ge, X., Ifenthaler, D., and Spector, J. (eds). (2015). *Emerging Technologies for STE(A)M Education: Full STE(A)M Ahead.* Dordrecht: Springer.

Greene, M. (1978). *Landscapes of Learning*. New York, NY: Teachers College Press.

Greene, M. (1998). *Releasing the Imagination: Essays on Education, the Arts and Social Change*. San Francisco, CA: Jossey-Bass.

Gu, L., Huang, R., and Marton, F. (2004). "Teaching with variation. A Chinese way of promoting effective mathematics learning," in *How Chinese Learn Mathematics: Perspectives from Insiders,* eds F. Lianghuo, W. Ngai-Ying, C. Jinfa, and L. Shiqi (Singapore: World Scientific Publishing).

Hammersley, M., Gomm, R., and Foster, P. (2009). "Case study and theory," in *Case Study Method*, eds R. Gomm, M. Hammersley, and P. Foster (London: Sage), 234–259.

Kontra, C., Lyons, D. J., Fischer, S. J., and Bellock, S. L. (2015). Physical experience enhances science learning. *Psychol. Sci.* 26, 737–749.

Korn-Bursztyn, C. (ed.). (2012). *Young Children and the Arts: Nurturing Imagination and Creativity*. Charlotte, NC: Information Age Publishing.

Leach, M., Rockstrom, J., Raskin, P., Scoones, I., Stirling, A., Smith, A., et al. (2012). Transforming innovation for sustainability. *Ecol. Soc.* 17:11.

Marshall, J. (2014). Transdisciplinarity and art integration: toward a new understanding of art-based learning across the curriculum. *Stud. Art Educ.* 55, 104–127.

Mendick, H., Berge, M., and Danielsson, A. (2017). A critique of the STEM pipeline: young people's identities in Sweden and science education policy. *Br. J. Educ. Stud.* 4, 1–17.

Russell-Bowie, D. (2009). Syntegration or disintegration? Models of integrating the arts across the primary curriculum. *Int. J. Educ. Arts* 10:28.

Sefton-Green, J., Thomson, P., Jones, K., and Bresler, L. (eds). (2011). *The Routledge International Handbook of Creative Learning.* Abingdon: Routledge.

Sterling, S. (2012). "Afterword: let's face the music and dance?," in *Learning for Sustainability in Times of Accelerating Change* eds A. E. J. Wals and P. B. Corcoran (Wageningen: Wageningen Academic Publishers), 511–516.

Stolz, S. (2015). Embodied learning. *Educ. Philos. Theory* 27, 474–487.

Trowsdale, J. (2016). Imagineering: recreating spaces through collaborative art-making. *Creat. Theor. Res. Appl.* 3, 274–291.

van Manen, M. (1990). *Researching Lived Experience.* New York, NY: SUNY Press.

van Boeckel, J. (2013). *At the Heart of Art and Earth: An Exploration of Practices in Arts-based Environmental Education.* Helsinki: Aalto University.

Wade-Leeuwen, B. (2016). *Out of the Shadows: Fostering Creativity in Teacher Education Programs.* Champaign, IL: Common Ground Publishing.

Warwick Commission Report (2015). *Enriching Britain: Culture, Creativity and Growth. The 2015 Report by the Warwick Commission on the Future of Cultural Value.* Available at: https://www2.warwick.ac.uk/research/warwickcommission/futureculture/finalreport/warwick_commission_report_2015.pdf [accessed February 23, 2017].

Westley, F., Scheffer, M., and Folke, C. (2012). Special issue: reconciling art and science for sustainability. *Ecol. Sci.* 17:11.

Woolery, L. A. (2006). *Art-based Perceptual Ecology as a Way of knowing the Language of Place*. Ph.D. dissertation, Antioch University, Yellow Springs, OH.

Index

About the Editors

Tatiana Chemi, Ph.D., is Associate Professor at Aalborg University, Chair of Educational Innovation, where she works in the field of artistic learning and creative processes. She is the author of many published articles and reports and is also the author of *Artbased Approaches. A Practical Handbook to Creativity at Work*, Fokus Forlag, 2006, *Kunsten at integrere kunst i undervisning* [The art of integrating the arts in education], Aalborg Universitetsforlag, 2012, *In the Beginning Was the Pun: Comedy and Humour in Samuel Beckett's Theatre*, Aalborg University Press, 2013 and *The Art of Arts Integration*, Aalborg University Press, 2014. In 2013, Aalborg University Press named her Author of the Year. Her latest work focuses on distributed creativity, artistic creativity and artistic partnerships published in the following contributions: with Jensen, J. B. & Hersted, L., *Behind the Scenes of Artistic Creativity*, Frankfurt, Peter Lang, 2015; "Distributed Problem-Solving: How Artists' Participatory Strategies Can Inspire Creativity in Higher Education." In Zhou, C. (Ed.). *Handbook of Research on Creative Problem-Solving Skill Development in Higher Education.* IGI global. 2016; "The Teaching Artist as Cultural Learning Entrepreneur: An Introductory Conceptualization." In *Teaching Artist Journal*. 2015. 13, 2, pp. 84–94. With Xiangyun Du she edited *Arts-based Methods and Organisational Learning: Higher Education Around the World*, soon to be published for the Palgrave Series Studies in Business, Arts and Humanities and she is preparing for the Palgrave Series Studies on Chinese Education in a Global Perspective, *Arts-based Education – China and its intersection with the world* (Du, Chemi & Wang, eds.). She is currently involved in research projects examining artistic creativity cross-culturally, arts-integrated educational designs in schools, theatre laboratory (publishing for Palgrave, *A Theatre Laboratory Approach to Pedagogy and Creativity: Odin Teatret and Group Learning*) and the role of emotions in learning.

Xiangyun Du, Ph.D., is a professor at Department of Learning and Philosophy, Aalborg University, and College of Education, Qatar University. Her main research interests include innovative teaching and learning in education, particularly, problem-based and project-based learning methods in fields ranging from engineering, medicine and health, and foreign language education, to diverse social, cultural and educational contexts. She has also engaged with educational institutions in over 10 countries in substantial work on pedagogy development in teaching and learning. Professor Du has over 140 relevant international publications including monographs, international journal papers, edited books and book chapters, as well as conference contributions. She has also been actively involved in a number of international academic programs, networks, and editorial works for journals. Currently she is also (co)editing book series for PalGrave and RIVER publishers.

About the Authors

Aimee Durning is a Senior Teaching Assistant at the University Primary School and a member of the Chartered College of Teaching. Aimee has ten years' primary school experience.

Alison Laurie Neilson is a senior researcher at the Centre for Social Studies, University of Coimbra. She has degrees in biology, environmental studies, and Comparative International Development Education. She conducts narrative and arts-informed research on the way sustainability is understood and manifest in education and policy. She studies the processes of education to understand how people are encultured into governance structures and how the issues and structures themselves are constructed.

Andrea Inocêncio is a visual artist and performer. She has received several grants and awards: Calouste Gulbenkian Foundation (Canada), Camões Institute (Argentina), KTH Royal Institute of Technology (Stockholm), INOV-ART/DGArtes (Barcelona), Eurodissée Program (Paris), DRAC (Azores), and the Federation Nationale de la Culture Française. She is co-founder of the performance collective Malparidas.

Arzu Mistry is an educator and artist. She teaches at the Srishti Institute for Art Design and Technology in Bangalore. Arzu maintains a high level of dedication and enthusiasm for the arts, as mediums for pedagogy, advocacy, transformation, and intervention for the building of sustainable inclusive communities. Arzu facilitates the Art in Transit and placeARTS public art projects in the city of Bangalore and internationally with the hope that art can facilitate dialogue between people and the urban spaces they inhabit. Arzu co-facilitates the Accordion Book Project and is the co-creator of the artist book Unfolding Practice: Reflections on Learning and Teaching. Her art and education practice connects teachers, youth and families with place using memory, story, play and art and design practices through inter- disciplinary education and public community art projects, livelihoods training, teacher professional development and educational research and practice.

Chiaki Ishiguro, Ph.D., is an Associate Researcher at Tamagawa University Brain Science Institute. She has been involved in research projects examining artistic creativity and inspiration, especially in education for cultivating creative fluency in university students. Her research is currently focused on artistic inspiration through art appreciation and the development of evaluation methods for art learning.

Dina Zoe Belluigi has recently joined Queen's University to coordinate a portfolio of international programmes in Higher Education Studies. She was previously a Senior Lecturer at CHERTL at Rhodes University, South Africa, where she coordinated both the Honours and Masters programmes, and taught on the philosophies and paradigms which construct notions of teaching and learning, the assessment of student learning, and the evaluation of teaching and courses. Dina has taught fine art studio practice and is a practicing artist. Current research interests include the politics of belonging in staff educational development; the relationship between practice-based learning, teaching and research; and the question of authorship and agency in higher education.

Hiroaki Ishiguro is a professor in the Department of Education at Rikkyo University. His scholarly interests focus on theoretical investigations of learning and development from a sociohistorical perspective. Dr. Ishiguro engages in qualitative, micro-analytic, video-based research. He has examined such phenomena as nursery care activities, language development, play activity, classroom discourse, schooling, and disability studies. More recently, he has focused on arts-based learning and performances in multicultural and multilingual contexts. He recently published an article titled 'How a young child learns how to take part in mealtimes in a Japanese day-care centre: a longitudinal case study' in the *European Journal of Psychology of Education* and a Japanese book titled *What do children learn in their classrooms? On research of learning practice* from the University of Tokyo Press.

James Biddulph is the first Headteacher of the University of Cambridge Primary School. He is currently completing his Ph.D. on creative learning in ethnic minority home contexts.

Kimber Andrews, Ph.D., spent her early career as a professional dancer, choreographer and dance educator before transitioning into scholarly research. Her research interests include: embodiment in education, artistry in teaching, and the relationship between Eastern philosophies and the practice of teaching. Her work rests at the intersection of many fields; hence she teaches in a variety of disciplines including dance, documentary filmmaking, qualitative research methodologies, and aesthetic education.

Kristóf Fenyvesi is a researcher of transdisciplinary STE(A)M learning at University of Jyväskylä. He is the founding director of the www.experienceworkshop.org international mathematics-art movement.

Luke Rolls is an Assistant Headteacher at the University of Cambridge Primary School. He collaborates with a number of schools as a Specialist Leader of Education to support the development of mathematics teaching.

Maja Maksimovic Assistant Professor at the Department for Pedagogy and Andragogy, University of Belgrade and a researcher at the Institute for Pedagogy and Andragogy. She is a deputy editor of the adult education journal studies the author of the numerous publications. Beside her experience in teaching and writing, her interests are related to performance art and theatre.

Margarida Augusto is a sociology student in the Faculty of Economics at the University of Coimbra. She did a internship at the Centre for Social Studies helping to develop and organize the network for arts informed research.

Maria Simões calls herself a multi-artist. She has been a clown, actor, director, teacher, producer, animator, music, writer, "playologist", and creator. She is the founder of Descalças-cultural cooperative (Açores). She likes to think that the world is becoming better every day. She likes to know that Art can transform the world and the lives of the people who live in it.

Nayla Naoufal is a postdoctoral fellow at the University of Oslo in Norway exploring the integration of ecocitizenship perspectives and artistic approaches. She holds a Ph.D. in Environmental Science and Education from Université du Québec à Montréal on the contribution of environmental education to the construction of peace. She has worked in art-based peace and sustainability education in museums and other contexts.

Pamela Burnard is Professor of Arts, Creativities and Education at the Faculty of Education, University of Cambridge.

Rannveig Björk Thorkelsdóttir is assistant Lecturer, at the University of Iceland, School of Education. She is an experienced drama teacher educator. She has been involved in curriculum development in creativity and introducing drama in compulsory schools and higher education. She is in her research and practice focused on drama and artistic approaches to teaching and learning. Rannveig has published several books and articles on teaching and learning in drama. She is a professional actress and she has written and directed plays for children and young adults.

Rigel Lazo Cantú has a Bachelor degree in Audiovisual Arts and a Master degree in Social Work, both by the Universidad Autónoma de Nuevo León in Monterrey, México. Has been working in community projects using street art not only as a tool to promote community development but community activism by democratising art and public space.

Rita São Marcos studied Communication Design in the Faculty of Fine Arts University of Lisbon and Sociology at the University of Coimbra. She has worked on governance of the environment in the Azorean islands. She works closely with Alison Neilson. For her Ph.D. she is exploring the tensions between participatory democracy, expert knowledge and small-scale fishers' participation in Common Fisheries Policy.

Rodrigo Lacerda studied Cinema and Television at London Metropolitan University and National Film and Television School and received his Master's degree in Anthropology, specializing in Visual Culture, from the New University of Lisbon. He has co-directed several documentaries and worked in the field of post-production for cinema and advertising in the United Kingdom and Portugal. He is pursuing his Ph.D. in Anthropology on the relationship between heritage and indigenous cinema in Brazil.

Simone Longo de Andrade is a consultant in human rights, mainly in international scenarios, with grassroots and social workers, developing creative, participatory and experiential learning methods to translate human rights into social change tools. Simone sees herself as a popular educator – education as the practice of freedom. She has worked with Dignity International on the human rights approach to development and poverty. She holds an Advanced

Law Degree (University of Coimbra) and a Master in Human Rights and Democratisation.

Susanne Jasilek is an artist and artist educator and formerly Artist-in-Residence at the Faculty of Education, University of Cambridge. She has been involved in ground-breaking programmes working in diverse settings with school children, families, artists, post graduates and the wider community.

Takeshi Okada, Ph.D., is a Professor at the Graduate School of Education of the University of Tokyo, where he teaches the psychology of creativity. Since obtaining a Ph.D. in Psychology at Carnegie Mellon University, he has studied the process of artistic creation and scientific discovery through field studies, psychological experiments and design-based research. His current research interests are in the process of artistic creation by artists and non-artists, the process of artists developing expertise, and ways of facilitating artistic creativity.

Tatjana Dragovic is a doctoral educator who works across different disciplines, sectors and industries and is recognized as an international educator and researcher.

Todd Elkin was born in Rutland VT and is a visual artist, writer, researcher, activist and arts educator currently living in Oakland, CA. He has an extensive visual arts background including work as a professional fine-art printmaker and as a freelance illustrator. He earned a BFA in Interdisciplinary Studies from the San Francisco Art Institute, and a masterŠs degree from the Harvard Graduate School of Education in the Arts In Education program. He is currently the Fine Arts Department Chair and an Art teacher at Washington High School in Fremont, CA where his specialty is designing and delivering student-driven, art-centered trans-disciplinary curriculum and fostering cultures of critical thinking and reflection. Promoting visual and media literacy are major goals in his teaching practice as well as encouraging students to see themselves as contemporary artists and global citizens taking part in conversations about important topics and issues. Elkin is a Senior Faculty member of The Integrated Learning Specialist Program offered through the Alameda County Office of Education. Since 2016 Elkin has been a mentor and presenter in the Art21 Educators Program. As an educator, Elkin is motivated by issues of social justice and equity in matters of race, class and gender and is committed to creating spaces for students to engage in relevant lines of inquiry about these and other important themes. Elkin is Co-Facilitator, with Arzu Mistry of The Accordion Book Project. In 2016 an artists' book he co-authored with Arzu Mistry, entitled Unfolding Practice: Reflections on Learning and Teaching was published by the Women's Studio Workshop. Elkin and Mistry are currently working on their second book.

Una MacGlone is a double bassist and lecturer at the Royal Conservatoire of Scotland, where she teaches free improvisation, community music techniques, supervises postgraduate students and coordinates the Healthy Musician module. Her research interests are in free improvisation, improvisation pedagogy, cross-disciplinary collaboration and Socio-Cultural approaches. She has given many workshops, lectures and presentations all over the UK as well as in North America and Europe.

Una is a founder member of the Glasgow Improvisers Orchestra, has improvised live on BBC Radio 3 and recorded many cds. She has been supported by the Thomas Laing Reilly Scholarship, from the University of Edinburgh for her PhD project, investigating improvisation with preschool children.

For Product Safety Concerns and Information please contact our EU representative GPSR@taylorandfrancis.com
Taylor & Francis Verlag GmbH, Kaufingerstraße 24, 80331 München, Germany

www.ingramcontent.com/pod-product-compliance
Ingram Content Group UK Ltd.
Pitfield, Milton Keynes, MK11 3LW, UK
UKHW022209190926
475702UK00003B/280
* 9 7 8 8 7 9 3 6 0 9 3 8 9 *